Hydraulics Research
Wallingford

Charts for the hydraulic design of channels and pipes

5th Edition

Charts for the hydraulic design of channels and pipes

Fifth edition

Published by Hydraulics Research,
Wallingford, Oxfordshire,
OX10 8BA, England.

Distributed by Thomas Telford Ltd
1–7 Great George Street
London SW1P 3AA

First published 1963
~~Fourth~~ edition 1983
Fifth

ISBN 0 946466 02 5

Printed in Great Britain by The Thetford Press Limited, Thetford, Norfolk

Foreword to First Edition

A multiplicity of formulae is available for the design of pipes and conduits, those of exponential form being the most used because of their ease of manipulation and availability in graphical form. In general, these formulae are empirical and have been derived from limited data; thus extrapolation can lead to serious error. Moreover, engineers have been confused by the differences between the various equations, and many are not clear as to which should be used in given circumstances, or the value to be given to their roughness coefficients.

The Colebrook-White equation, which was developed in 1939 from the Prandtl-Karman theory of turbulence, supersedes these purely empirical formulae, and has been shown by many experimenters to be applicable to virtually any commercial surface and fluid, over wide ranges of conditions. This equation is now recognised as the best available means of estimating friction loss, but unfortunately its complexity has prevented it coming into general use.

Since its derivation almost twenty years ago, several attempts have been made to simplify its application by expressing it in graphical form. None of these has succeeded in reducing the equation to the simple form required by engineers: that is, as design charts expressed directly in terms of discharge, hydraulic gradient, pipe or channel dimensions, and velocity.

The problem has been under review at the Hydraulics Research Station, under my direction, for some time, and I was convinced that a solution in chart or tabular form was feasible. The difficulties were finally overcome by a combination of dimensional analysis and graphical methods leading to a series of design charts in terms of the four engineering variables.

These design charts — the Wallingford charts — are contained in this publication. With them a direct solution can be obtained to almost any problem of fluid resistance, utilizing the best available equation for friction loss.

C C INGLIS
Director of Hydraulics Research 1947-58

Foreword to Fifth Edition

Since publication of the first edition of Hydraulics Research Paper No. 2 there have been considerable developments in pipe manufacturing processes and jointing procedures; new pipe materials have also been introduced. New roughness data, based largely on HRS tests, have therefore been incorporated in this edition together with new information on slimed sewers.

This fifth edition is the responsibility of Dr W R White, Group Manager, Overseas Projects. Mrs P M Brown and Dr R Bettess supervised the preparation of the text and the charts.

Dr T J WEARE
Managing Director

Hydraulics Research Station Limited
Wallingford, Oxfordshire
1983

Contents

Page

Introduction 1

The Colebrook-White equation 1

The linear measure of surface roughness 3

Explanation of the charts and tables 3

(1) The design charts, Figs 1 to 17 3

(2) The shape coefficient P/R for non-circular sections, Fig 18 4

(3) Proportional velocity and discharge in part-full circular sections, Fig 19 4

(4) Correction to the design charts for variation in temperature, Table 2 5

(5) Recommended roughness values, Table 3 5

Surfaces of composite roughness 5

Examples 6

1) Use of design charts for pipes, Figs 1 to 13 6

2) Use of design charts for channels, Figs 3 to 17 6

3) Use of the shape coefficient graphs, Fig 18 7

4) Proportional velocity and discharge in pipes flowing part-full, Fig 19 8

5) Correcting the design charts for temperatures other than 15°C, Table 2 8

References 9

Nomenclature 9

Contents (Cont'd) **Page**

Tables

Table 1:	Limiting velocities for the region of turbulent flow (based on 1 per cent differences)	11
Table 2:	Velocity corrections for variation in temperature	11
Table 3:	Recommended roughness values	12

Figures

Fig 1:	k_s = 0.003mm	16
Fig 2:	k_s = 0.006mm	17
Fig 3:	k_s = 0.015mm	18
Fig 4:	k_s = 0.03mm	20
Fig 5:	k_s = 0.06mm	22
Fig 6:	k_s = 0.15mm	24
Fig 7:	k_s = 0.3mm	26
Fig 8:	k_s = 0.6mm	28
Fig 9:	k_s = 1.5mm	30
Fig 10:	k_s = 3.0mm	32
Fig 11:	k_s = 6.0mm	34
Fig 12:	k_s = 15mm	36
Fig 13:	k_s = 30mm	38
Fig 14:	k_s = 60mm	40
Fig 15:	k_s = 150mm	41
Fig 16:	k_s = 300mm	42
Fig 17:	k_s = 600mm	43
Fig 18:	Shape coefficient P/R for non-circular sections	44
Fig 19:	Proportional velocity and discharge in part-full circular sections	45
Table 3:	Repeated from page 12	46

Charts for the hydraulic design of channels and pipes

Introduction

One of the earliest attempts at providing a scientific basis of channel design was Chezy's work about 1775. In the latter half of the last century and in the early years of this century, an appreciation of the shortcomings of the Chezy equation led to the development of many empirical formulae, some of which have remained in use to this day. Exponential forms such as the Manning equation have been the most popular because of the ease of manipulation and because they can readily be expressed in graphic form as design charts. However, most empirical formulae are based on limited data, and, as in general they have no sound physical basis, extrapolation outside the range of experimental confirmation may lead to serious error.

When the need for dimensional homogeneity was appreciated, and non-dimensional parameters such as Reynolds number were available to represent the relationships in problems containing many variables, the physics of fluid friction became more apparent. In the 1930's, Karman and Prandtl published theories of turbulent flow, which, coupled with experiments on smooth pipes (Stanton and Pannell) and on artificially roughened pipes (Nikuradse), provided the foundations for major advances in design techniques.

These theories were used as a basis by Colebrook and White in their derivation of an equation describing the frictional resistance of pipes. Ref 1 details the background to the Colebrook-White equation. To simplify the application of this formula, tables and charts of solutions have been prepared by the Hydraulics Research Station. Design charts are reproduced in this publication together with an account of turbulent flow and the application of the equation to design. Explanations of the charts and associated tables are included with worked examples. Tables for the hydraulic design of pipes are published in Ref 2.

The Colebrook-White equation

The Prandtl-Karman theory of turbulence led to the following flow equations:

Smooth turbulence:

$$1/\sqrt{\lambda} = 2\log \mathcal{R}\sqrt{\lambda} - 0.8 \qquad \text{1}$$

Rough turbulence:

$$1/\sqrt{\lambda} = 2\log D/2k_s + 1.74 \qquad \text{2}$$

where

λ = friction coefficient, $2gDS/V^2$

k_s = a linear measure of effective roughness

D = pipe diameter

$\mathcal{R}$ = Reynolds number, VD/ν

A rough pipe follows equation 2 at high Reynolds numbers when viscous effects are negligible, and a 'square' law of resistance applies, ie V is proportional to $\sqrt{S}$. The same pipe, at lower velocities or carrying a more viscous fluid would deviate further from equation 2 as Reynolds number reduced. Provided its roughness was not extreme, it would tend to act increasingly like a smooth pipe, approaching equation 1. The roughness of a pipe can be expressed as the ratio R/k_s, where R is the hydraulic mean depth (area/wetted perimeter). The smaller the value of R/k_s then the lower is the velocity required for smooth turbulence. There is a gradual transition between rough-turbulence and smooth-turbulence which depends to some extent upon the character of the surface. In this region roughness and viscosity operate jointly in determining resistance over a range of conditions. Most commercial surfaces follow a very similar transition curve, and the equation obtained by Colebrook is:

Transitional flow:

$$\frac{1}{\sqrt{\lambda}} = -2 \log \left[\frac{k_s}{3.7D} + \frac{2.51}{\Re\sqrt{\lambda}}\right] \quad \ldots\ldots 3$$

The limiting conditions within each turbulence range are tabulated in Table 1.

The transition equation incorporates the smooth-turbulent and rough-turbulent formulae. It is thus applicable to virtually any surface roughness, conduit size, velocity of flow, or fluid. Its advantage over purely empirical expressions lies in this wide range of application.

If the non-dimensional variables of equation 3 are replaced by their dimensional engineering equivalents, the following expresion results:

$$V = -2\sqrt{(2gDS)} \log \left[\frac{k_s}{3.7D} + \frac{2.51\nu}{D\sqrt{(2gDS)}}\right] \quad \ldots\ldots 4$$

where

V = velocity

g = gravitational acceleration

S = hydraulic gradient; invert slope for full pipes, water surface slope for open channels

ν = kinematic viscosity of fluid

For open channel flow, this may be expressed in terms of the hydraulic mean depth:

$$V = -\sqrt{(32gRS)} \log \left[\frac{k_s}{14.8R} + \frac{1.255\nu}{R\sqrt{(32gRS)}}\right] \quad \ldots\ldots 5$$

The intractability of the Colebrook-White equation in its ordinary forms is apparent: R and S occur under the initial root sign and, in different combinations, within the logarithm bracket. Unless both the size and gradient are the stated variables, a direct solution is impossible and a trial-and-error method becomes necessary. To avoid the labour of calculation the charts published here reduce the equation to the four engineering variables; discharge, gradient, size and velocity.

The various forms of the transition equation have been stated here so that they are available for direct calculation. However, the design charts, Figs 1-17, will enable most problems to be solved without further reference to the equation.

For circular sections flowing full the relation of discharge to conduit size and average velocity of flow is simply stated as $Q = \frac{1}{4}\pi D^2 V$. This permits a scale of discharge to be shown on a flow chart for pipes, having diameter and velocity as its logarithmic axes. With non-circular sections, area is not expressed solely in terms of the hydraulic mean depth, but a convenient discharge equation is obtained by using the ratio of perimeter to hydraulic mean depth as a shape coefficient, thus:

$$4\pi QR/P = 4\pi R^2 V \quad \ldots\ldots 6$$

Although a discharge scale cannot be shown directly on an open-channel flow chart, $4\pi QR/P$ may take its place, and a simple calculation will convert this to discharge for a channel of any predetermined shape. A subsidiary diagram relating discharge, P/R and $4\pi RQ/P$, has been added to the design charts which does this graphically, and the shape coefficient P/R has also been worked out for several common channel shapes, see Fig 18.

When flow conditions are near the rough turbulent region an allowance for viscous effects can be made by an approximation to the transition equation as follows:

$$V = 4\sqrt{2gS\,k_s} \,.\, \sqrt{R/k_s} \,.\, 14.8R/k_s - 8\nu/k_s \quad \cdots\cdots 7$$

The limiting velocities to which these approximations may be applied are shown in Table 1.

The linear measure of surface roughness

The pipes used by Nikuradse in his experiments were roughened by means of a uniformly-graded sand glued to the surface. The diameter of the sand grains provided a standard of comparison against which other surfaces may be judged, their roughness thus being quoted as the size of sand which would give the same resistance under rough turbulent flow. Much experimental data has been re-examined on these lines, so that k_s values are available for most surfaces. Table 3 summarises the recommended design values, and covers most types of construction materials.

The Colebrook-White equation applies in particular to roughnesses consisting of separate protuberances of random height and spacing. Most commercial surfaces are of this form. Some classes of pipe have a quite different form of roughncss, eg corrugated metal culverts, and do not follow the same resistance function in the transition zone between smooth turbulence and rough turbulence. Such surfaces are not included in the list.

The k_s values bear some relation to the physical dimension of the roughness projections, and therefore a visual examination of a particular surface will give a guide to its roughness. Strickler's investigation of natural channels indicated that the k_s value corresponded to the size which was exceeded by 10 per cent of the exposed bed material. In other engineering fields, the larger of the roughness elements probably have a similar predominating effect.

Explanation of the charts and tables

(1) The design charts, Figs 1 to 17

Design charts have been prepared corresponding to a series of roughness values of commercial surfaces, carrying water at a temperature of 15°C.

Most of the charts, Figs 3 to 13, are given in pairs — left hand page for circular pipes and the right hand page for non-circular sections. The complete set of charts, Figs 1 to 17, relates to seventeen roughness values from k_s = 0.003, 0.006, 0.015, etc to 150, 300, 600mm, and the respective categories of construction are listed on each pipe chart. The velocity range covered is from 0.06m/s to 6m/s, with pipe sizes from 0.025m to 2.5m in diameter, and channels of hydraulic mean depth from 0.06 to 6m.

The four engineering variables appear on each chart, so that if any two are known a point is defined on the chart which solves the two unknowns. With non-circular sections of known shape, the coefficient P/R permits discharge to be read directly from the upper chart (see Examples 1 and 2).

(2) The shape coefficient P/R for non-circular sections, Fig 18

This coefficient occurs in the relationship of discharge to velocity and hydraulic mean depth (equation 6). Values are readily calculated for the common channel shapes, and expressed in terms of the ratio of bed width to depth of flow. This has been done for rectangular and trapezoidal channels of several side slopes. The resultant graphs are plotted together with similar data for pipes flowing part full.

If the ratio of bed width to depth is known, reference to this chart will give a corresponding value of P/R to be used in evaluating discharge (see Example 3).

(3) Proportional velocity and discharge in part-full circular sections, Fig 19

Sometimes, having designed a gravity pipeline for a certain discharge flowing full, the depth of flow for some lesser discharge is required. Alternatively, the designer may wish to know the velocities and discharges at particular depths of flow within the pipe. These problems are readily solved using the chart giving proportional velocity and discharge in terms of proportional depth. As the Colebrook-White transition formula includes the effects of viscosity as well as roughness, the proportionality relationships are rather complicated and include logarithmic terms. The equations are:

Rough-turbulent:

$$V_p = [1 + \frac{\log R_p}{\log (3.7\ D/k_s)}] \sqrt{R_p} \quad \cdots\cdots 8$$

Smooth turbulent:

$$V_p = [1 + \frac{\log R_p}{\log (0.54D\ (2gS/\nu^2)^{1/3})}] \sqrt{R_p} \quad \cdots\cdots 9$$

These equations, being of similar form, give a family of similar curves. The parameter to be used for the rough-turbulent region is defined by the value of D/k_s, whilst the corresponding parameter for smooth-turbulence is $\theta = 0.146D\ (2gS/\nu^2)^{1/3}$.

The proportionality expression can be generalised to include the transition region, thus:

$$V_p = [1 + \frac{\log R_p}{\log (3.7\theta)}] \sqrt{R_p} \quad \cdots\cdots 10$$

where

$$\theta \simeq [\frac{k_s}{D} + \frac{1}{3600\ DS^{1/3}}]^{-1} \text{ (m units, water at 15°C)} \quad \cdots\cdots 11$$

In most practical problems, one of the two terms in the bracket will be unimportant.

The use of the proportional velocity and discharge chart is demonstrated by Example 4.

(4) Correction to the design charts for variation in temperature, Table 2

The dimensioned charts refer to water at 15°C, and though it would be quite straightforward to prepare similar sets of charts corresponding to the amended viscosity at other temperatures, a more satisfactory method is to make an arithmetic correction for the effect of a change of temperature. This may be appreciable at small k_s values, but will be unnecessary for very rough surfaces.

Expressing the correction to velocity (m/s) as

$$V = V_{15} + at_{15} \qquad \cdots\cdots 12$$

where t_{15} = actual temperature — 15°C, the coefficient a depends on the roughness and the product RS. Values have been tabulated which, with some interpolation give temperature adjustment. The use of this table is shown in Example 5.

(5) Recommended roughness values, Table 3

The table of surface roughnesses is based on recommendations by Colebrook(3), Rouse(4), King(5), Lamont(6) and Perkins and Gardiner(7). The surfaces listed are classified as good, normal or poor examples of their respective categories, thus leaving to the engineer's judgement the actual value to be used in any particular scheme. The range of roughness covered by good and poor, takes account not only of the quality of the jointing but also the variation in surface roughness to be found in pipes that are normally of the same material. This list is not intended to absolve him of the responsibility for checking, by precise hydraulic tests whenever possible, the actual surface roughness achieved in particular projects. Where such direct evidence is available, from comparable projects, it should clearly take precedence over the general roughness values quoted here.

Surfaces of composite roughness

If the wetted perimeter of a channel or pipe is composed of surfaces of dissimilar roughness, for example where sliming has taken place only below the usual water level or where the invert is of concrete and the walls of brick, there are two ways of computing discharges using the charts in this report.

The first method is based on the concept of an equivalent grain roughness for the whole perimeter and utilises the expression

$$k_s = p_1 k_{s1} + p_2 k_{s2}$$

where p_1, p_2 denote the proportions of the total perimeter occupied by surfaces 1 and 2, and k_{s1}, k_{s2} denote the equivalent sand grain roughness of surfaces 1 and 2.

This method may be used in situations where the difference in roughness values is not excessive and where the two surfaces occupy similar proportions of the total perimeter. It will also give approximate answers outside these ranges. These ranges can be defined (somewhat arbitrarily) as:

$$0.05 < k_{s1}/k_{s2} < 20 \text{ and } 0.33 < p_1/p_2 < 3.0$$

The advantage of the above method is that it provides a direct solution to the problem of composite roughness and the charts can be used without the need to resort to successive approximation techniques.

The second method is based on the concept of an equivalent friction factor for the whole perimeter and utilises the expression

$\lambda_s = p_1 \lambda_1 + p_2 \lambda_2$

where the suffices denote surfaces 1 and 2 as before.

The method given below is not restricted to the limited ranges of conditions of the first method but, even so, errors will occur where there are large differences in the grain roughnesses of the two surfaces, say $k_{s1}/k_{s2} < 0.01$ or $k_{s1}/k_{s2} > 100$. The method requires a successive approximation approach as follows:

(i) Determine an approximate value for k_s using:

$k_s = p_1 k_{s1} + p_2 k_{s2}$

(ii) Use the appropriate chart to determine the flow, Q

(iii) Calculate $\mathcal{R}$

(iv) Calculate $C_1 = \dfrac{k_{s1}}{14.8R} + \dfrac{5.1286}{\mathcal{R}^{0.89}}$

(v) Calculate $C_2 = \dfrac{k_{s2}}{14.8R} + \dfrac{5.1286}{\mathcal{R}^{0.89}}$

(vi) Determine C from: $\log C = \dfrac{\log C_1}{[p_1 + (\log C_1/\log C_2)^2\, p_2]^{1/2}}$

(vii) Calculate $k_s = 14.8R\,(C - \dfrac{5.1286}{\mathcal{R}^{0.89}})$

(viii) Use the appropriate chart to determine the flow, Q

(ix) Repeat (iii) to (viii) until Q changes by less than the required tolerance

Examples 1) *Use of design charts for pipes, Figs 1 to 13*

A spun precast concrete pipe is required to carry a storm water discharge of 950 l/s when laid at a gradient of 1 in 430. Calculate its size.

Data:

Q 950 l/s

S = 0.00232 = 0.232 m per 100m

k_s = 0.6mm (see Table 3)

Fluid, water at average temperature.

On the design chart for k_s = 0.6 mm, read Q = 950 l/s on the top scale and 0.232m per 100m on the right hand scale. The intersection of these inclined grids gives

V = 1.49 m/s (left hand scale)

D = 0.90m (bottom scale)

A 900mm diameter pipe or near equivalent is therefore required.

2) *Use of design charts for channels, Figs 3 to 17*

(a) A stream diversion channel with 45° side-slopes is to be lined with concrete, using rough timber formwork. A bed width to depth ratio of 3 to 1 has been decided on. The available gradient is 1 in 1050 and its capacity is to be 15 m³/s. Calculate the design width.

Data:

Q $= 15m^3/s$

S $= 0.00095 = 9.5m$ per 10 000m

k_s $= 1.5mm$

Fluid, water at average temperature

P $= (3 + 2\sqrt{2})d$

A $= 4d^2$

R $= A/P = 0.686d$

P/R $= 8.50$

Using the upper chart of Fig 9 the vertical through $Q = 15m^3/s$ and the horizontal through $P/R = 8.5$ intersect at $4\pi QR/P = 22.1$ which is transferred down the sloping grid to the lower chart. The intersection of this line with 9.5m per 10 000m on the hydraulic gradient grid (sloping, slightly-curved lines) gives the required unknowns from the other two scales.

V $= 2.06m/s$

R $= 0.925m$

As $R = 0.686d$ for this sectional shape, $d = 1.35m$ and hence $B = 4.05m$. Calculate $Q = AV$ as a check: $Q = 4 \times 1.82 \times 2.06 = 15.0m^3/s$.

(b) When the above channel is carrying $10m^3/s$, what will be the depth of flow?

As the shape of the waterway, in terms of its bed width to depth ratio, is not known in advance, a method of successive approximation is required. The following method converges very rapidly.

Step 1 Assume $P/R = 8.5$ as for the channel when full. Enter the upper chart with $P/R = 8.5$ and $Q = 10m^3/s$. Transfer to the main chart, ($4\pi QR/P = 15$), and read off V at the intersection with the hydraulic gradient of 9.5 in 10 000, viz. $V = 1.88m/s$. Hence $A = Q/V = 5.32m^2$.

But from the channel's geometry,

$A = (4.05 + d)\ d\ m^2$

Inserting $A = 5.32$ gives

$d^2 + 4.05d - 5.32 = 0$

Solving this gives $d = 1.05m$ and hence $P = 7.05m$, and $P/R = 9.3$.

Step 2 Repeat the procedure with $P/R = 9.3$, $Q = 10m^3/s$. This yields $V = 1.82m/s$, $A = 5.49m^2$, $d = 1.07m$, $P = 7.08m$, and $P/R = 9.1$.

Step 3 Repeat with this value of P/R, giving $V = 1.84m/s$, $A = 5.44m^2$, $d = 1.07m$. The first approximation was correct to within about 2 per cent, so that generally one or two steps will suffice. As a general rule, where there is no advance indication of the probable P/R value, start the calculation at $P/R = 10$.

3) Use of the shape coefficient graphs, Fig 18

What is the value of P/R for an open channel of side-slope 1 in 2, with bed width of 40m and a depth of 8m?

For a ratio B/d of 5, the appropriate curve of Fig 18 gives directly, $P/R = 12.8$.

4) *Proportional velocity and discharge in pipes flowing part full, Fig 19*

Calculate the velocity and depth of flow in a 1.0m diameter spun concrete pipe for a discharge of 500 l/s. The slope of the pipe is 0.00232 and the surface roughness is 0.6mm.

Data:

D = 1.0m

S = 0.00232

Fluid, water at average temperature

Q_o = 1260 l/s } from the design chart for k_s = 0.6mm
V_o = 1.60 m/s }

Q = 500 l/s

The proportionality curves of Fig 19 are marked with values of θ, where

$$\theta = [\frac{k_s}{D} + \frac{1}{3600\ DS^{1/3}}]^{-1}$$

In this example,

$k_s/D = 0.60 \times 10^{-3}$

$1/3600\ DS^{1/3} = 2.09 \times 10^{-3}$

and

$\theta = 370$

Using the plotted curve for θ = 1000, the proportional discharge of 0.40 is read on the upper scale, and the left hand scale yields

Proportional depth = 0.45

d = 0.45 x 1 = 0.45m

The proportional velocity curve then yields

V = 0.95 x 1.60 = 1.52m/s

(In most examples of rough construction, $\theta \simeq D/k_s$)

5) *Correcting the design charts for temperatures other than 15°C, Table 2*

A 100mm diameter asbestos-cement water main carries 3.50 l/s at 15°C. What will it carry at the same hydraulic gradient at 2°C and at 25°C.

Data:

D = 100 mm

Q = 3.50 l/s

S = 0.235 m in 100m } from the design chart for k_s = 0.015 mm
V = 0.45 m/s }

Referring to Table 2, with k_s = 0.015mm and DS = 0.2mm, a = 0.0014

At 2°C, ie $t_{15} = -13$

V = 0.45 + 0.0014 x (−13)

= 0.45 − 0.018 = 0.432m/s

Thus Q = AV = 3.39 l/s at 2°C = 97% of discharge at 15°C

At 25°C, ie $t_{15} = +10$

V = 0.45 + 0.014 = 0.464

Thus Q = AV = 3.64 l/s at 25°C = 104% of discharge at 15°C.

References

1. ACKERS P. Resistance of fluids flowing in channels and pipes, Hydraulics Research Paper No. 1, HMSO, 1958.
2. HYDRAULICS RESEARCH WALLINGFORD. Tables for the hydraulic design of pipes and sewers. Fourth edition, 1983.
3. COLEBROOK C F. Turbulent flow in pipes, with particular reference to the transition region between the smooth and rough pipe laws. J. Instn. Civ. Engrs. 1939, **11**, 133.
4. ROUSE H. Evaluation of boundary roughness, Proc 2nd Hydraulic Conf., University of Iowa, 1943, Bulletin 27.
5. KING H W. Handbook of hydraulics, McGraw Hill, New York, 1954, Section 6.
6. LAMONT P A. A review of pipe friction data and formulae, with a proposed set of exponential formulae based on the theory of roughness, Proc. Instn. Civ. Engrs., Part 3, 1954, **3**, 248.
7. PERKINS J A and GARDINER I M. The effect of sewage slime on the hydraulic roughness of pipes. Report IT 218, Hydraulics Research Station, Wallingford, 1982.

Nomenclature

A	Area of flow cross-section
B	Width of channel at bed
a	Temperature correction coefficient
C′	Coefficient in $V = C'R^{\alpha}S^{\beta}$
D	Pipe diameter
d	Depth of water
g	Gravitational acceleration
h	Head loss
k_s	Linear measure of roughness, the Nikuradse equivalent sand roughness
log	Common logarithm (base 10)
N	Laminar flow shape factor
o	Suffix denoting pipe full
P	Wetted perimeter of flow section
p	Suffix denoting proportional value
Q	Discharge (volume per unit time)
R	Hydraulic mean depth, A/P
$\mathcal{R}$	Reynolds number, VD/ν or $4VR/\nu$
S	Hydraulic gradient, head loss per unit length
t	Temperature (°C)
t_{15}	Temperature difference relative to 15°C, t − 15°C
V	Average flow velocity Q/A

α,β Exponents in $V = C'R^{\alpha}S^{\beta}$

ν Kinematic viscosity

θ Coefficient for part full pipes in proportional flow equations

λ Friction factor (Darcy-Weisbach), $\lambda = 2gDS/V^2$

Table 1 **Limiting velocities in m/s for the regions of turbulent flow (based on 1 per cent differences)**

	Roughness, k_s mm				
	30	**3.0**	**0.3**	**0.03**	**0.003**
Rough turbulence at velocities exceeding:	0.03	0.3	3.0	30	—
Transitional flow, $R/k_s > 250$, in velocity range:	< 0.03	< 0.3	0.01 -3.0	0.1 - 30	> 1
Approximation to transition range at velocities exceeding:	0.006	0.06	0.6	6	—
Smooth turbulence, at velocities less than, for roughness ratios					
R/k_s = 10 000	—	—	0.03	0.3	3.0
R/k_s = 1000	—	—	0.015	0.15	1.5

Table 2 **Velocity corrections for variation in temperature**

$V = V_{15} + at_{15}$ (in m/s) where $t_{15} = t - 15$ (in degrees centigrade)

RS (mm)	DS (mm)	Roughness value, k_s in mm							
		0.006	**0.015**	**0.03**	**0.06**	**0.15**	**0.30**	**0.60**	**1.5**
500	2000	0.0327	0.0151	0.0080	0.0041	0.0017	0.0008	0.0004	0.0002
200	800	0.0290	0.0142	0.0077	0.0040	0.0017	0.0008	0.0004	0.0002
100	400	0.0257	0.0134	0.0075	0.0040	0.0016	0.0008	0.0004	0.0002
50	200	0.0222	0.0124	0.0071	0.0039	0.0016	0.0008	0.0004	0.0002
20	80	0.0174	0.0107	0.0065	0.0037	0.0016	0.0008	0.0004	0.0002
10	40	0.0140	0.0093	0.0060	0.0035	0.0016	0.0008	0.0004	0.0002
5	20	0.0110	0.0079	0.0054	0.0033	0.0015	0.0008	0.0004	0.0002
2	8	0.0077	0.0060	0.0044	0.0029	0.0014	0.0008	0.0004	0.0002
1	4	0.0057	0.0048	0.0037	0.0026	0.0013	0.0008	0.0004	0.0002
0.5	2	0.0042	0.0037	0.0030	0.0022	0.0012	0.0007	0.0004	0.0002
0.2	0.8	0.0028	0.0025	0.0022	0.0017	0.0011	0.0007	0.0004	0.0002
0.1	0.4	0.0020	0.0019	0.0017	0.0014	0.0009	0.0006	0.0004	0.0002
0.05	0.2	0.0014	0.0014	0.0013	0.0011	0.0008	0.0005	0.0003	0.0002
0.02	0.08	0.0009	0.0009	0.0008	0.0008	0.0006	0.0004	0.0003	0.0002
0.01	0.04	0.0007	0.0006	0.0006	0.0006	0.0005	0.0004	0.0003	0.0002

The values of 'a' tabulated above show the velocity adjustment in m/s for a 1°C temperature difference

Table 3 Recommended roughness values

Classification (assumed clean and new unless otherwise stated)	Suitable values of k_s (mm) Good	Normal	Poor
Smooth materials			
Drawn non-ferrous pipes of aluminium, brass, copper lead etc, and non-metallic pipes of Alkathene, glass, perspex etc	—	0.003	—
Asbestos-cement	0.015	0.03	—
Metal			
Spun bitumen or concrete lined	—	0.03	—
Wrought iron	0.03	0.06	0.15
Rusty wrought iron	0.15	0.6	3.0
Uncoated steel	0.015	0.03	0.06
Coated steel	0.03	0.06	0.15
Galvanised iron, coated cast iron	0.06	0.15	0.3
Uncoated cast iron	0.15	0.3	0.6
Tate relined pipes	0.15	0.3	0.6
Old tuberculated water mains with the following degrees of attack:			
Slight	0.6	1.5	3.0
Moderate	1.5	3.0	6.0
Appreciable	6.0	15	30
Severe	15	30	60
(Good: Up to 20 years use, Normal: 40 to 50 years use, Poor: 80 to 100 years use)			
Wood			
Wood stave pipes, planed plank conduits	0.3	0.6	1.5
Concrete			
Precast concrete pipes with 'O' ring joints	0.06	0.15	0.6
Spun precast concrete pipes with 'O' ring joints	0.06	0.15	0.3
Monolithic construction against steel forms	0.3	0.6	1.5
Monolithic construction against rough forms	0.6	1.5	—
Clayware			
Glazed or unglazed pipe:			
With sleeve joints	0.03	0.06	0.15
With spigot and socket joints and 'O' ring seals — dia < 150mm	—	0.03	—
With spigot and socket joints and 'O' ring seals — dia > 150mm	—	0.06	—
Pitch fibre (lower value refers to full bore flow)	0.003	0.03	—
Glass fibre	—	0.06	—
uPVC			
With chemically cemented joints	—	0.03	—
With spigot and socket joints, 'O' ring seals at 6 to 9 metre intervals	—	0.06	—
Brickwork			
Glazed	0.6	1.5	3.0
Well pointed	1.5	3.0	6.0
Old, in need of pointing	—	15	30

Table 3 (Cont'd)

Classification (assumed clean and new unless otherwise stated)	Suitable values of k_s (mm) Good	Normal	Poor
Slimed sewers*			
Sewers slimed to about half depth; velocity, when flowing half full, approximately 0.75m/s:			
Concrete, spun or vertically cast	—	3.0	6.0
Asbestos cement	—	3.0	6.0
Clayware	—	1.5	3.0
uPVC	—	0.6	1.5
Sewers slimed to about half depth; velocity, when flowing half full, approximately 1.2m/s:			
Concrete, spun or vertically cast	—	1.5	3.0
Asbestos cement	—	0.6	1.5
Clayware	—	0.3	0.6
uPVC	—	0.15	0.3
Sewer rising mains All materials, operating as follows			
Mean velocity 1m/s	0.15	0.3	0.6
Mean velocity 1.5m/s	0.06	0.15	0.30
Mean velocity 2m/s	0.03	0.06	0.15
Unlined rock tunnels			
Granite and other homogeneous rocks	60	150	300
Diagonally bedded slates (values to be used with *design* diameter)	—	300	600
Earth channels			
Straight uniform artificial channels	15	60	150
Straight natural channels, free from shoals, boulders and weeds	150	300	600

* The roughness of a slimed sewer varies considerably during any year. The normal value is that roughness which is exceeded for approximately half of the time. The poor value is that which is exceeded, generally on a continuous basis, for one month of the year. The value of k_s should be interpolated for velocities between 0.75 and 1.2m/s.

Table 3 is repeated at the end of the publication

Figures

List of surfaces with roughness k_s = 0·003 mm

Good examples of	Pitch fibre pipes running full bore
Normal examples of	Drawn non-ferrous pipes of aluminium, brass, copper, lead, etc and non metallic pipes of alkathene, glass, perspex, etc.
Poor examples of	

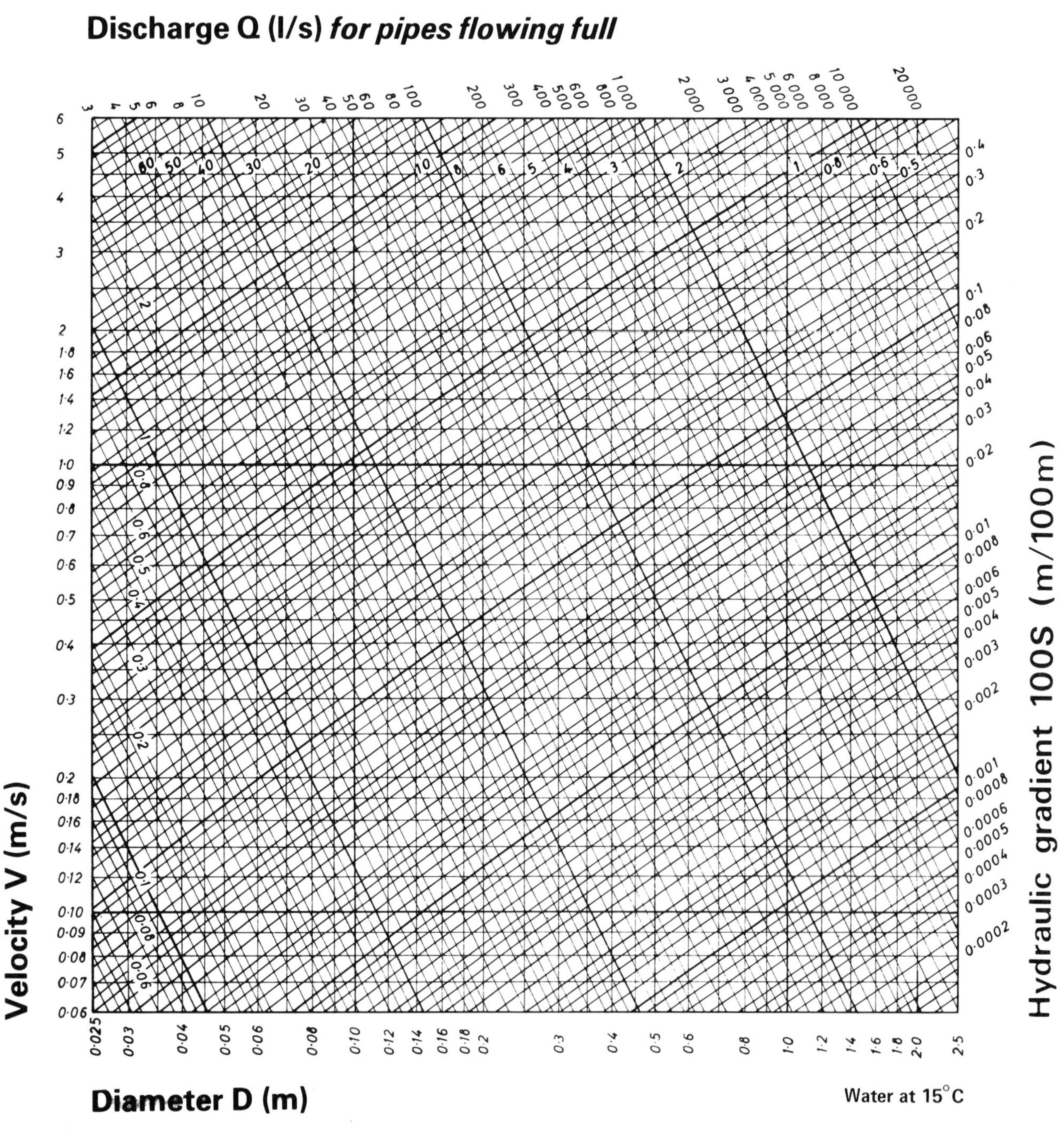

Fig 1 k_s = 0·003 mm

List of surfaces with roughness k_s = 0·006 mm

Good examples of

Normal examples of

Poor examples of

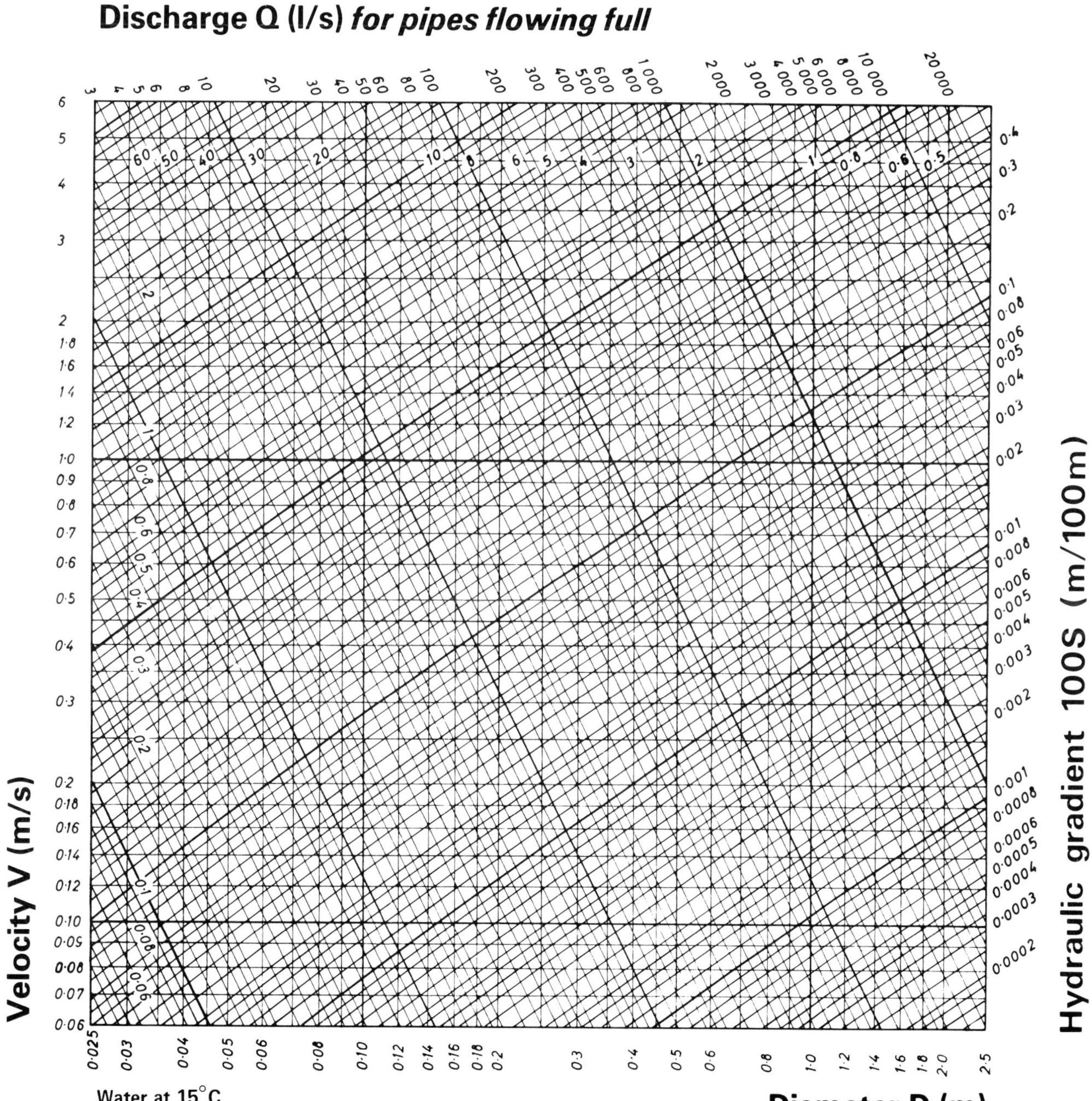

k_s = 0·006 mm Fig 2

List of surfaces with roughness k_s = 0·015 mm

Good examples of	Asbestos cement, uncoated steel
Normal examples of	
Poor examples of	

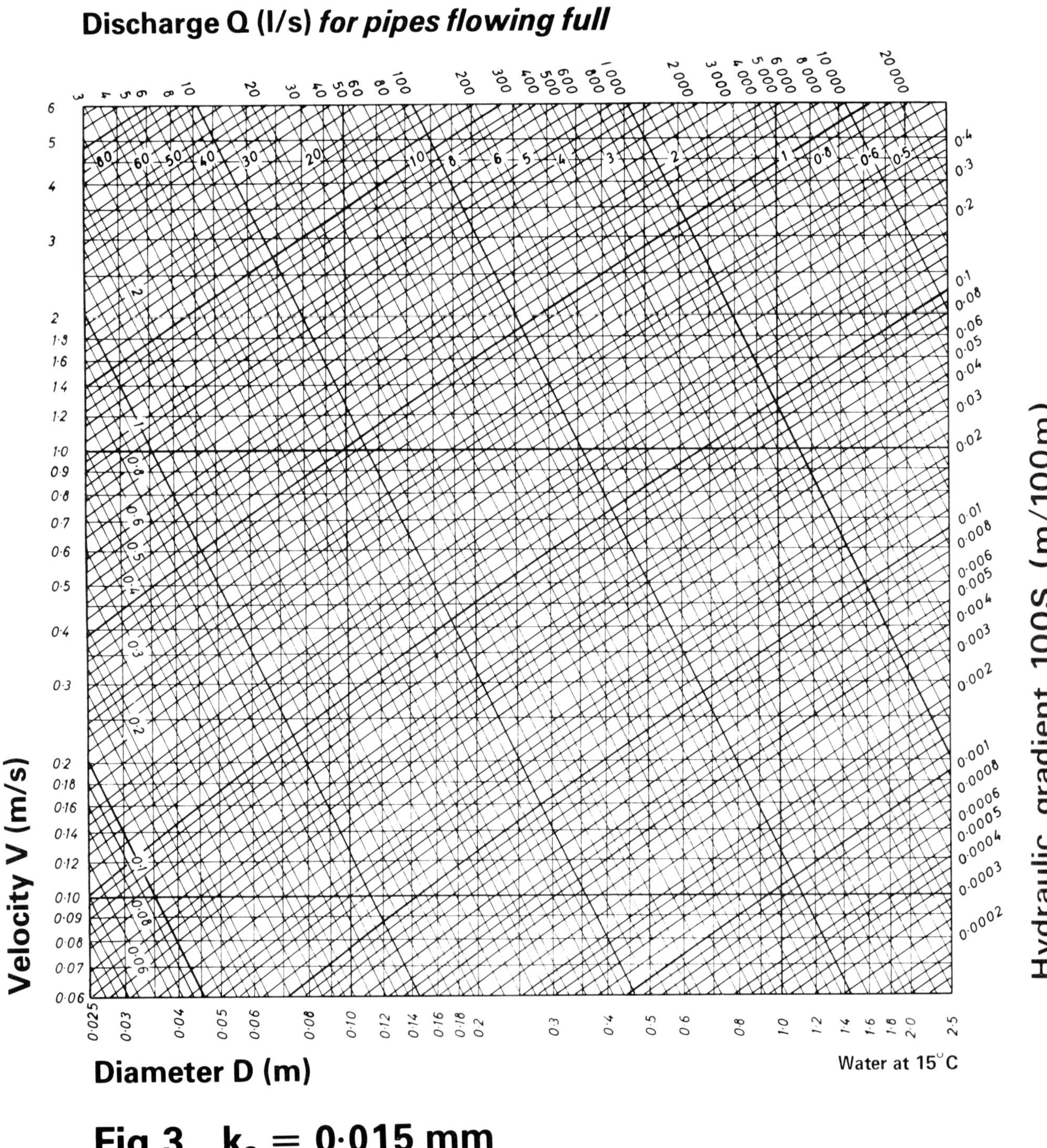

Fig 3 k_s = 0·015 mm

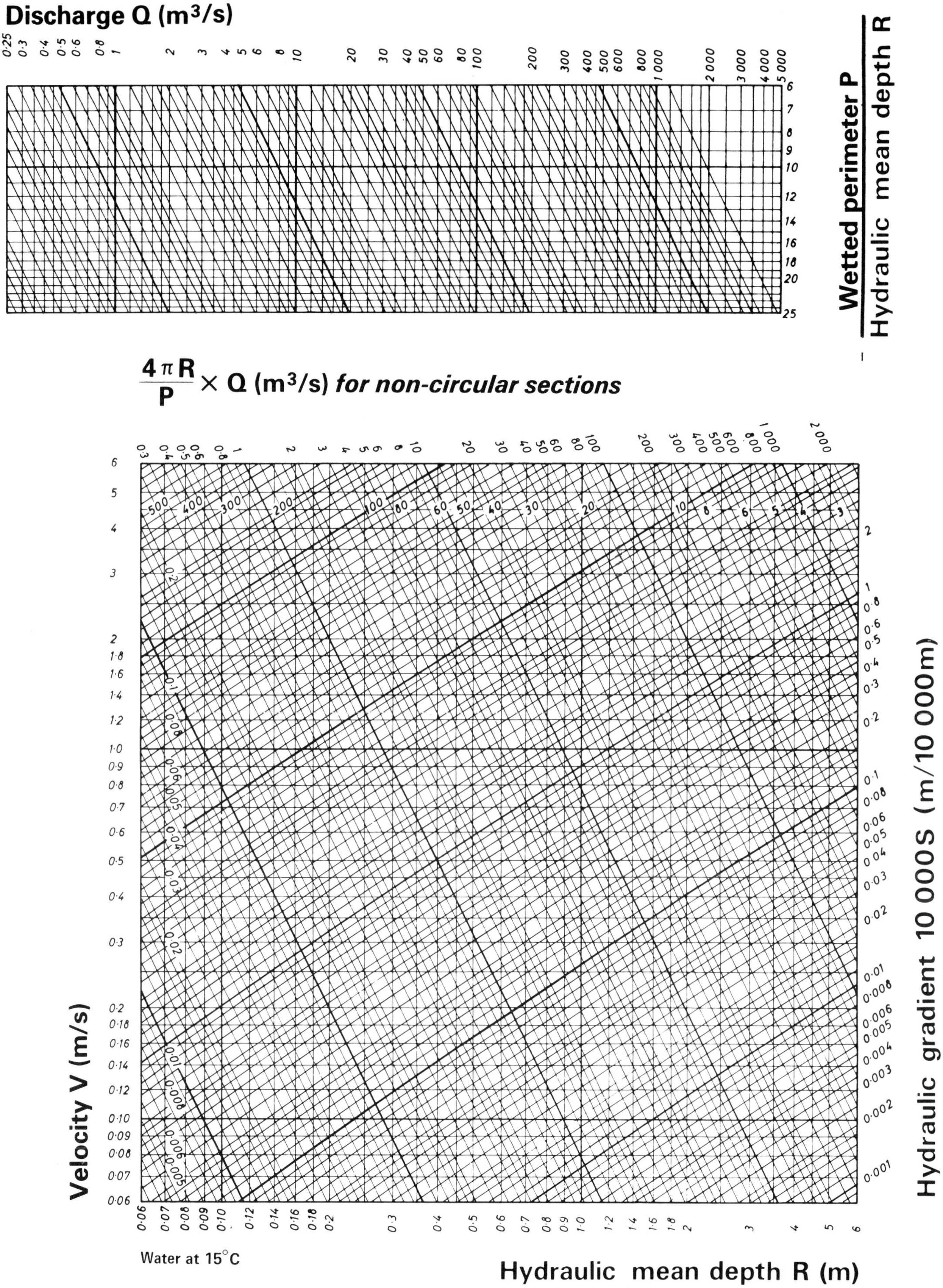

$k_s = 0{\cdot}015$ mm Fig 3

List of surfaces with roughness k_s = 0.03 mm

Good examples of	Wrought iron Coated steel Clayware (glazed or unglazed) with sleeve joints Sewer rising mains, mean velocity 2 m/s
Normal examples of	Asbestos cement Spun bitumen lined metal pipes Spun concrete lined metal pipes Uncoated steel Clayware (glazed or unglazed) with spigot and socket joints and 'O' ring seals – dia <150 mm Pitch fibre pipes running part full uPVC with chemically cemented joints
Poor examples of	

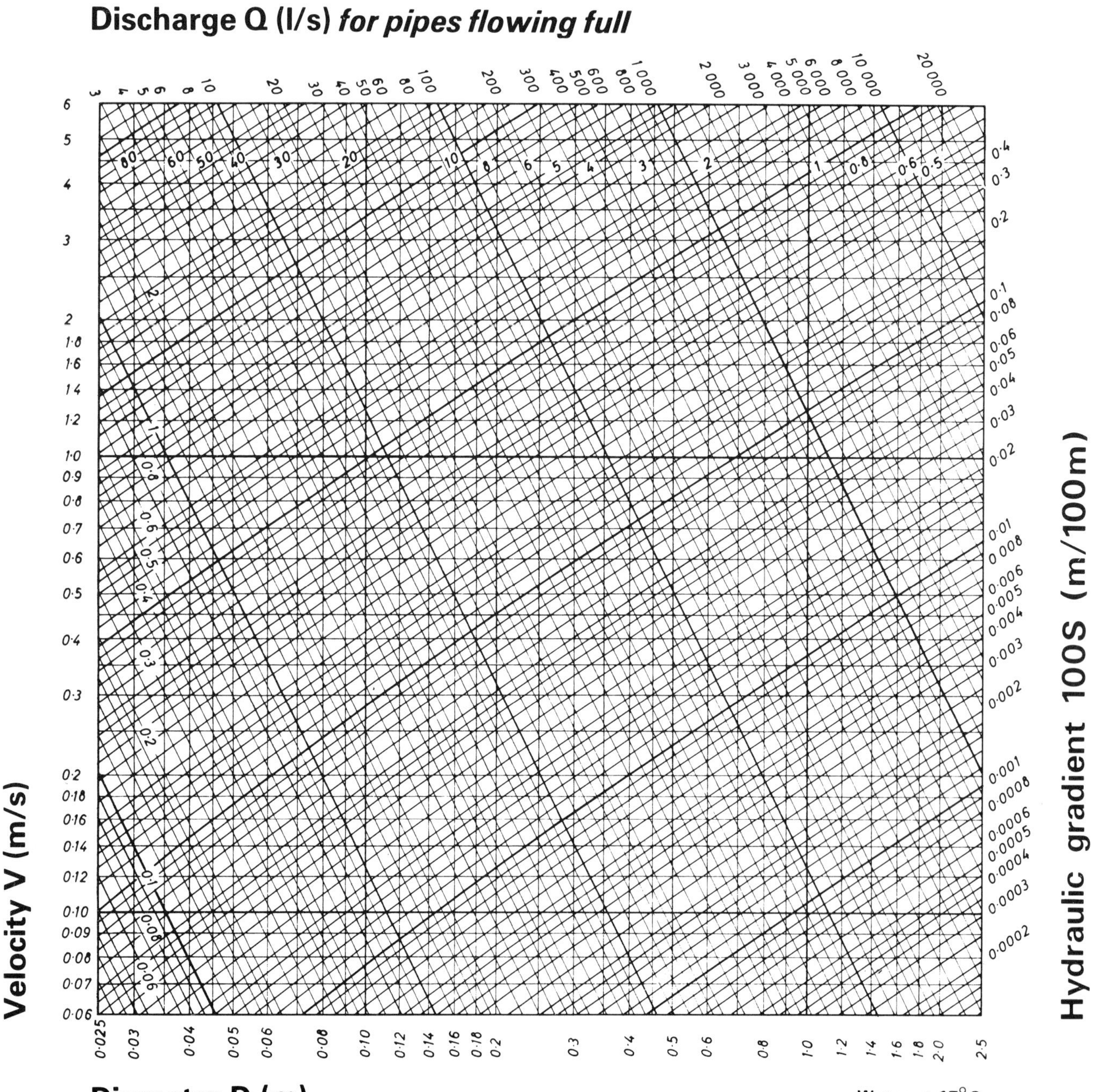

Fig 4 k_s = 0·03 mm

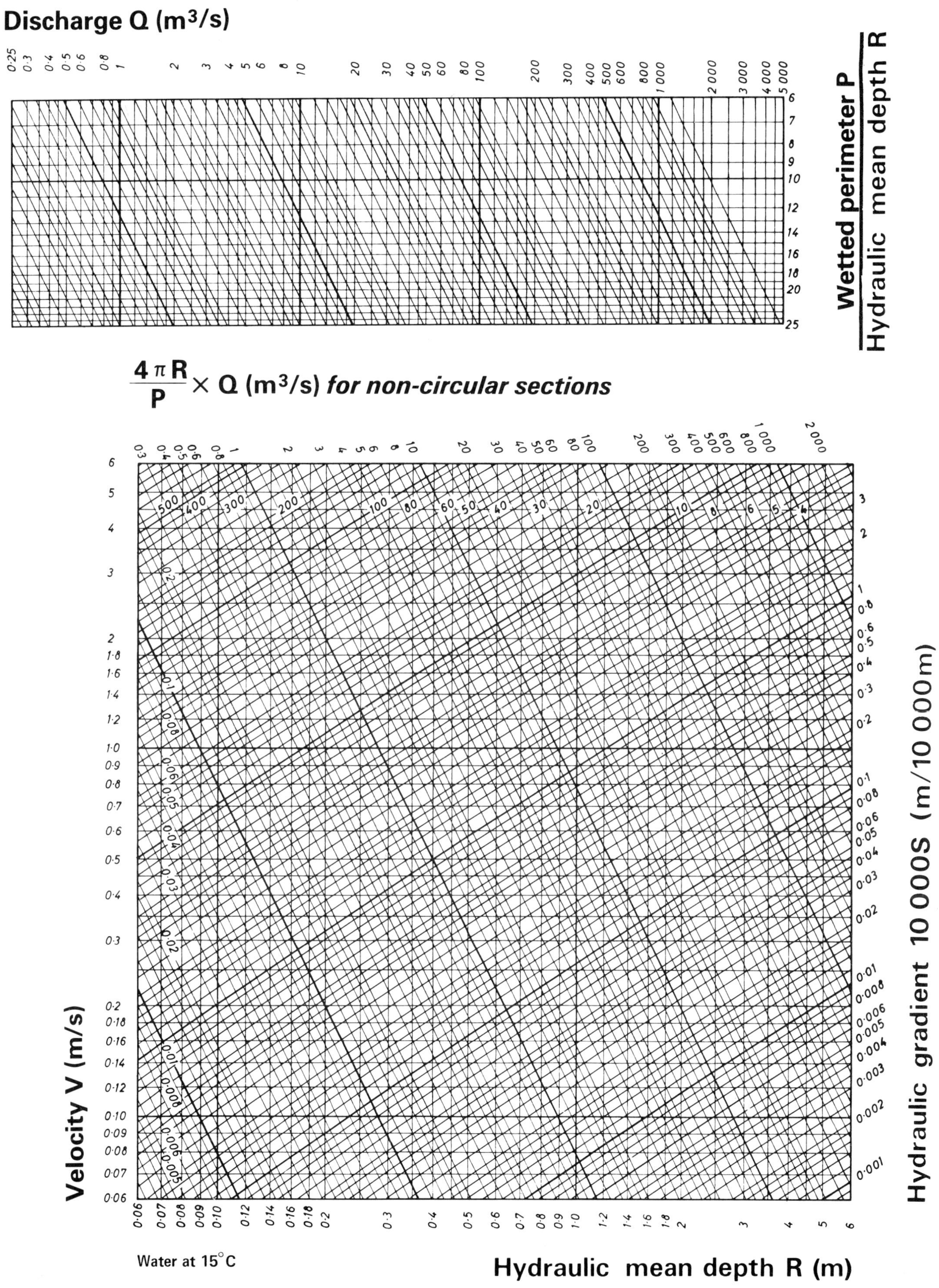

k_s = 0·03 mm Fig 4

List of surfaces with roughness k_s = 0.06 mm

Good examples of
- Galvanised iron, coated cast iron
- Precast concrete pipes with 'O' ring joints
- Spun precast concrete pipes with 'O' ring joints
- Sewer rising mains, mean velocity 1.5 m/s

Normal examples of
- Wrought iron
- Coated steel
- Clayware (glazed or unglazed) with sleeve joints
- Clayware (glazed or unglazed) with spigot and socket joints and 'O' ring seals – dia >150 mm
- Glass fibre
- uPVC with spigot and socket joints, 'O' ring seals at 6m to 9m intervals
- Sewer rising mains, mean velocity 2 m/s

Poor examples of
- Uncoated steel

Discharge Q (l/s) ***for pipes flowing full***

Velocity V (m/s)

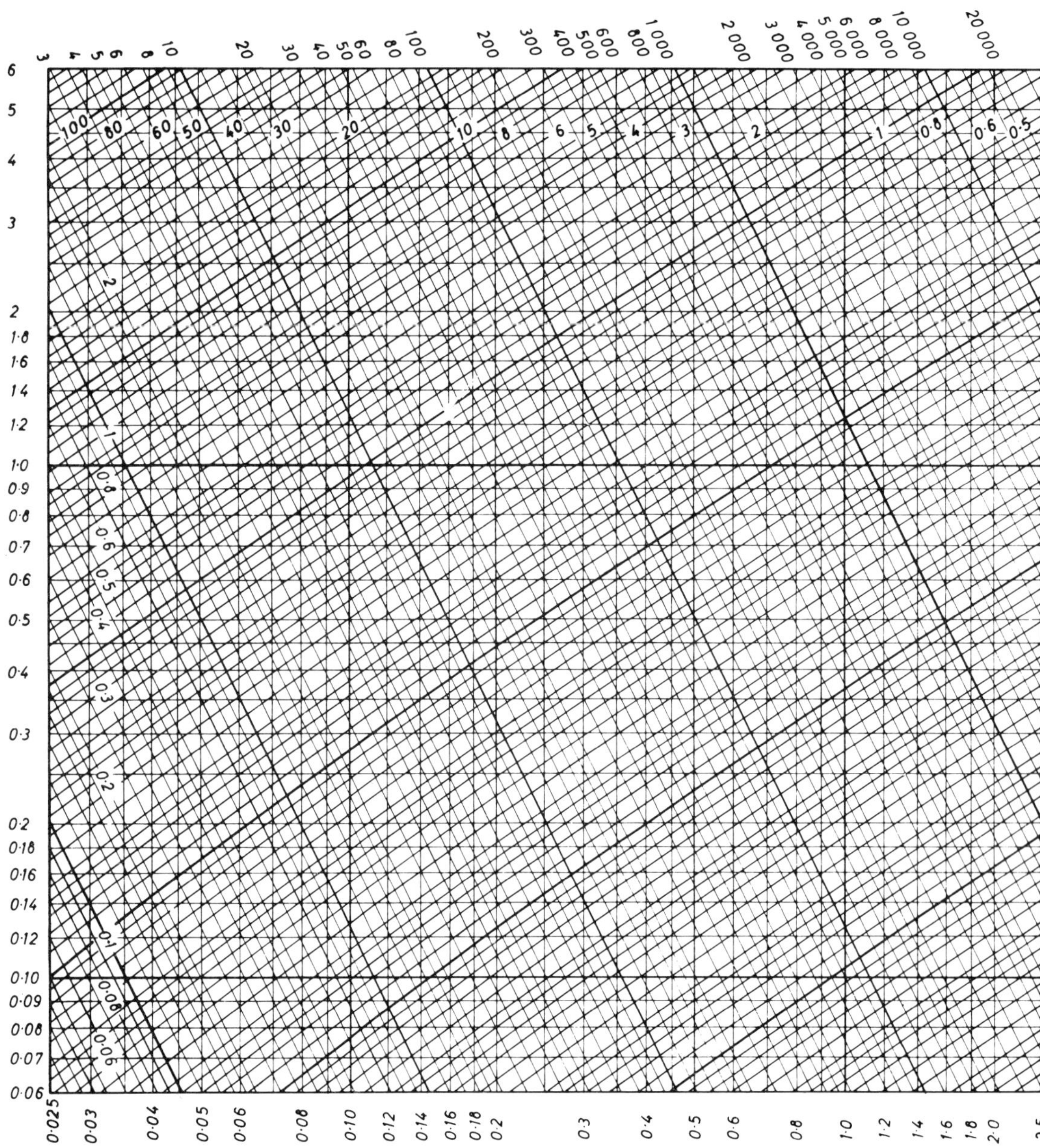

Hydraulic gradient 100S (m/100m)

Diameter D (m)

Water at 15°C

Fig 5 k_s = 0·06 mm

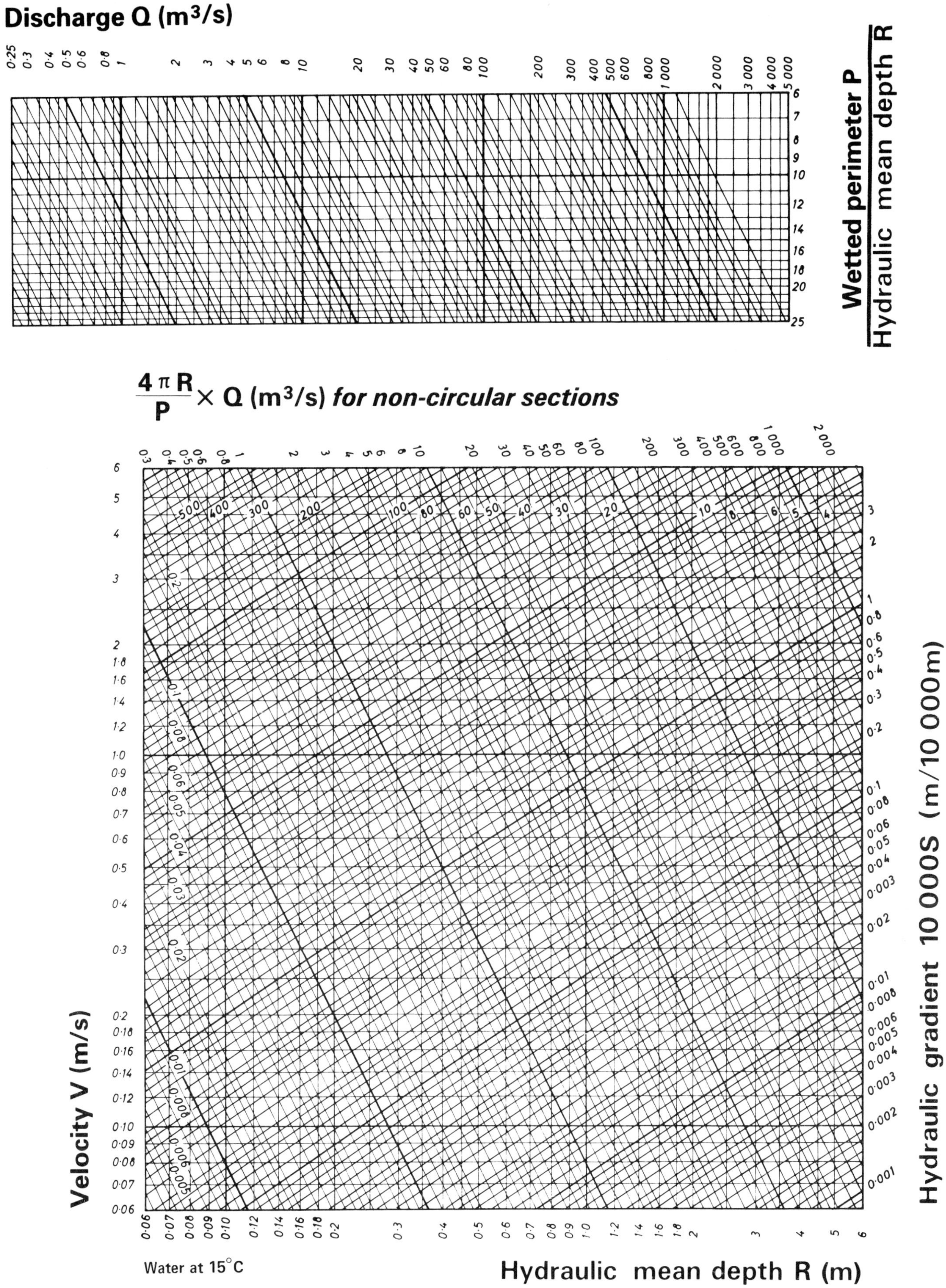

$k_s = 0{\cdot}06$ mm   **Fig 5**

List of surfaces with roughness k_s = 0.15 mm

Good examples of	Rusty wrought iron
	Uncoated cast iron
	Tate relined pipes
	Sewer rising mains, mean velocity 1 m/s
Normal examples of	Galvanised iron
	Coated cast iron
	Precast concrete pipes with 'O' ring joints
	Spun precast concrete pipes with 'O' ring joints
	uPVC sewers slimed to about half depth; velocity when flowing half full approximately 1.2 m/s
	Sewer rising mains, mean velocity 1.5 m/s
Poor examples of	Wrought iron
	Coated steel
	Clayware (glazed or unglazed) with sleeve joints
	Sewer rising mains, mean velocity 2 m/s

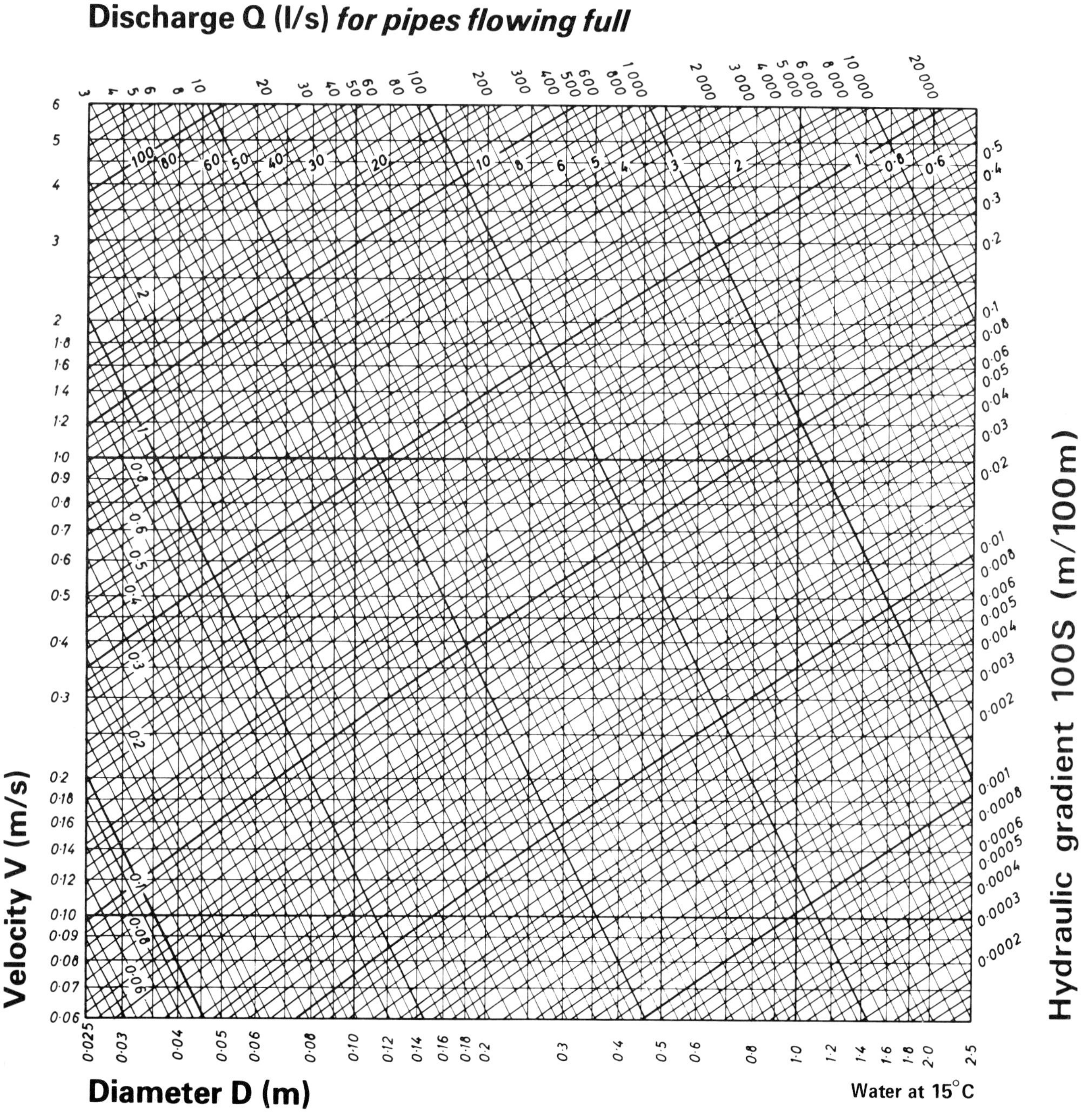

Fig 6 k_s = 0·15 mm

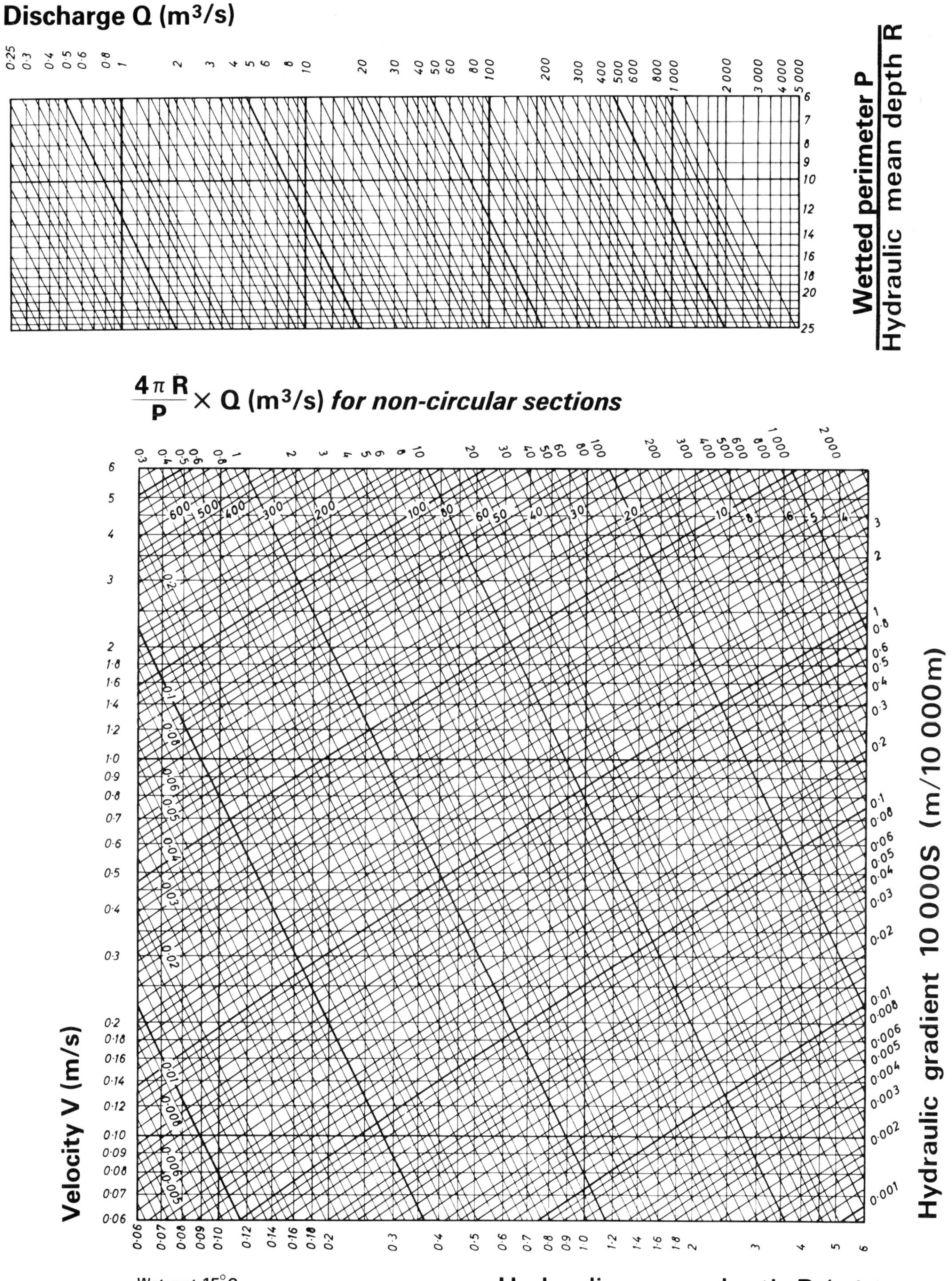

k_s = 0·15 mm Fig 6

List of surfaces with roughness k_s = 0.3 mm

Good examples of	Wood stave pipes, planed plank conduits Concrete monolithic construction against steel forms
Normal examples of	Uncoated cast iron Tate relined pipes Clayware sewers slimed to about half depth; velocity, when flowing half full, approximately 1.2m/s Sewer rising mains, mean velocity 1 m/s
Poor examples of	Galvanised iron, coated cast iron Spun precast concrete pipes with 'O' ring joints uPVC sewers slimed to about half depth; velocity when flowing half full approximately 1.2 m/s Sewer rising mains, mean velocity 2 m/s

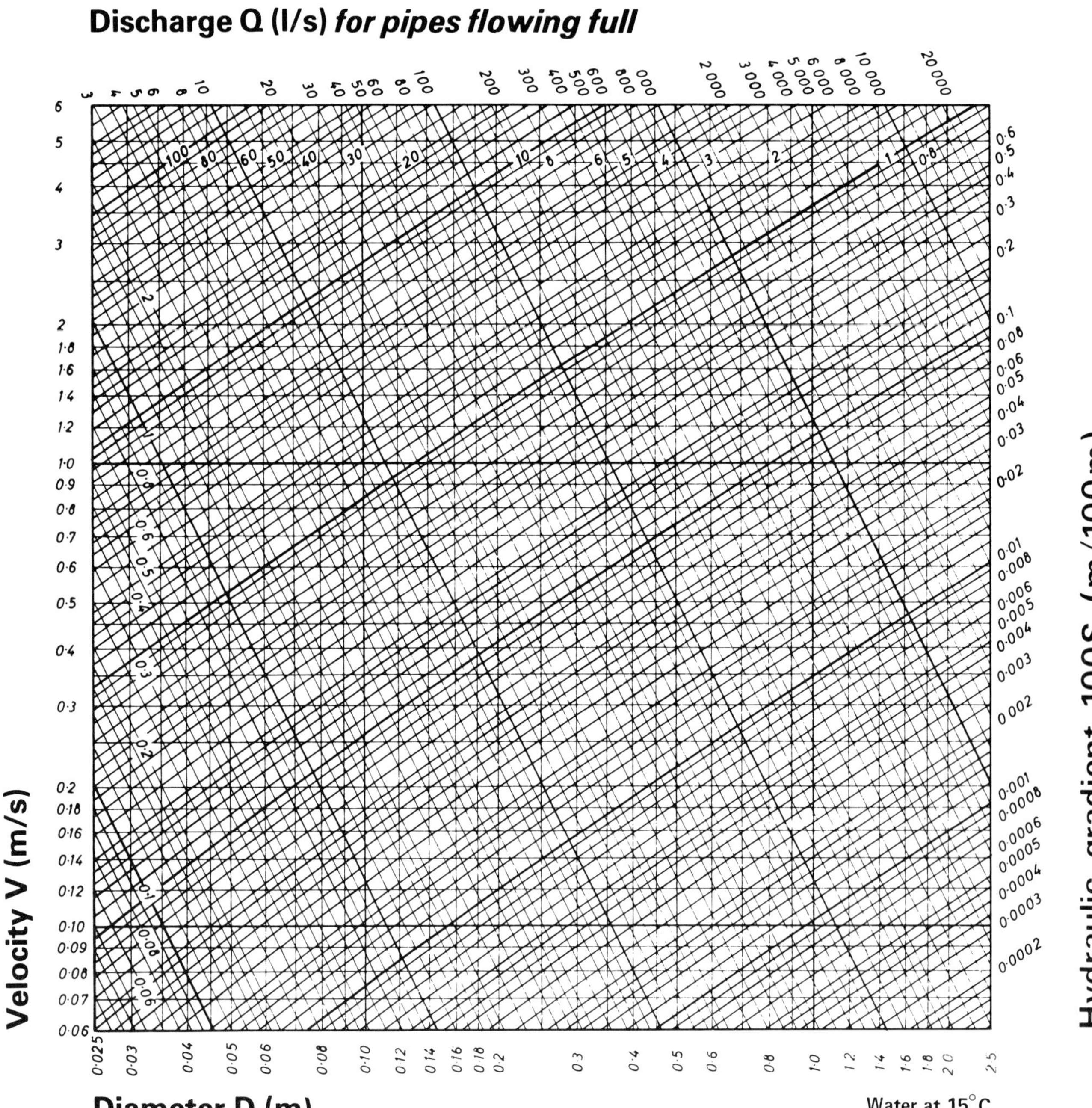

Fig 7 k_s = 0·3 mm

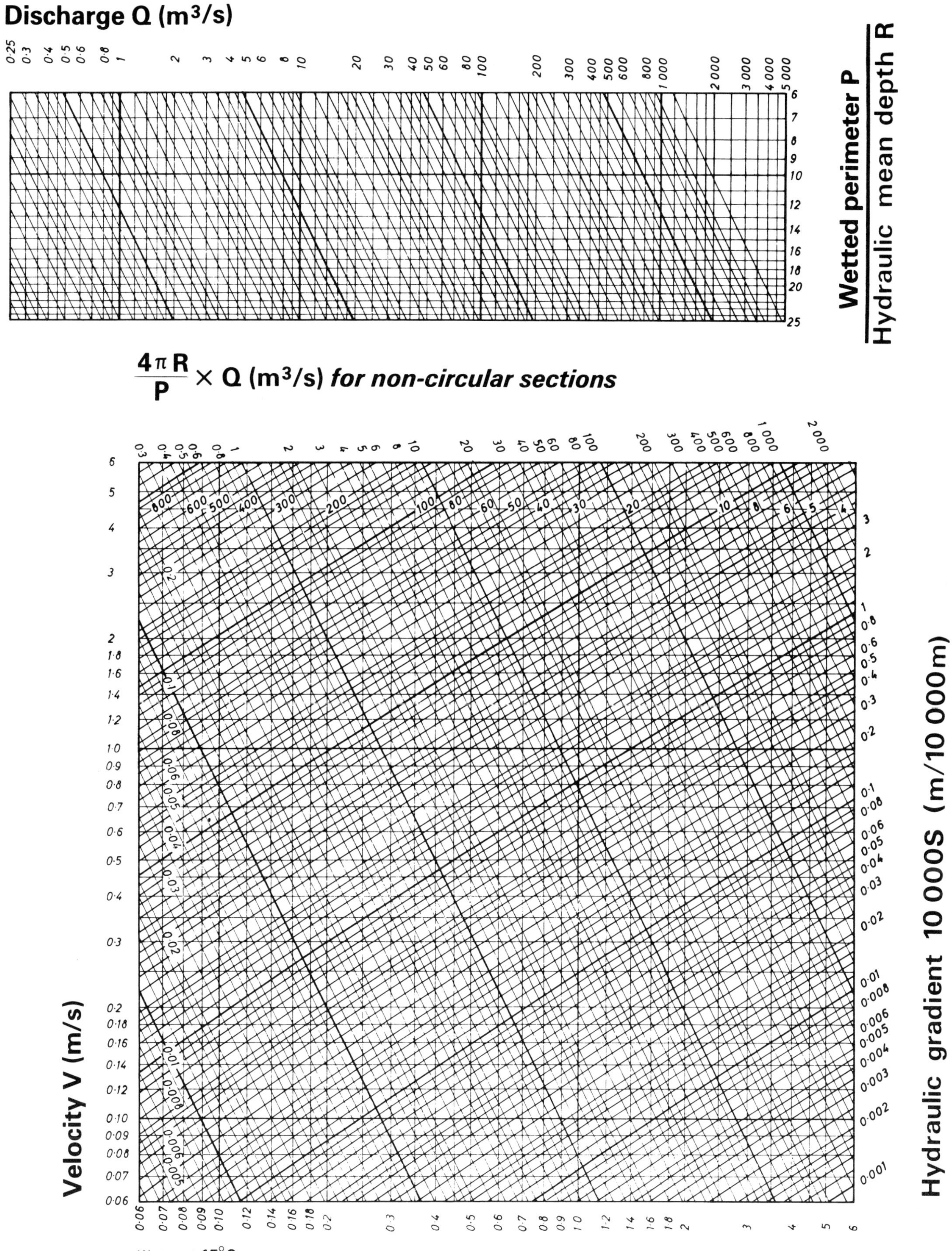

$k_s = 0{\cdot}3$ mm   Fig 7

List of surfaces with roughness k_s = 0.6 mm

Good examples of	Old tuberculated water mains with slight attack Concrete pipes, monolithic construction against rough forms Glazed brickwork
Normal examples of	Rusty wrought iron Wood stave pipes, planed plank conduits Concrete pipes, monolithic construction against steel forms uPVC sewers slimed to about half depth; velocity when flowing half full approximately 0.75 m/s Asbestos cement sewers slimed to about half depth; velocity, when flowing half full, approximately 1.2m/s
Poor examples of	Uncoated cast iron Tate relined pipes Precast concrete pipes with 'O' ring joints Clayware sewers slimed to about half depth; velocity, when flowing half full, approximately 1.2m/s Sewer rising mains, mean velocity 1 m/s

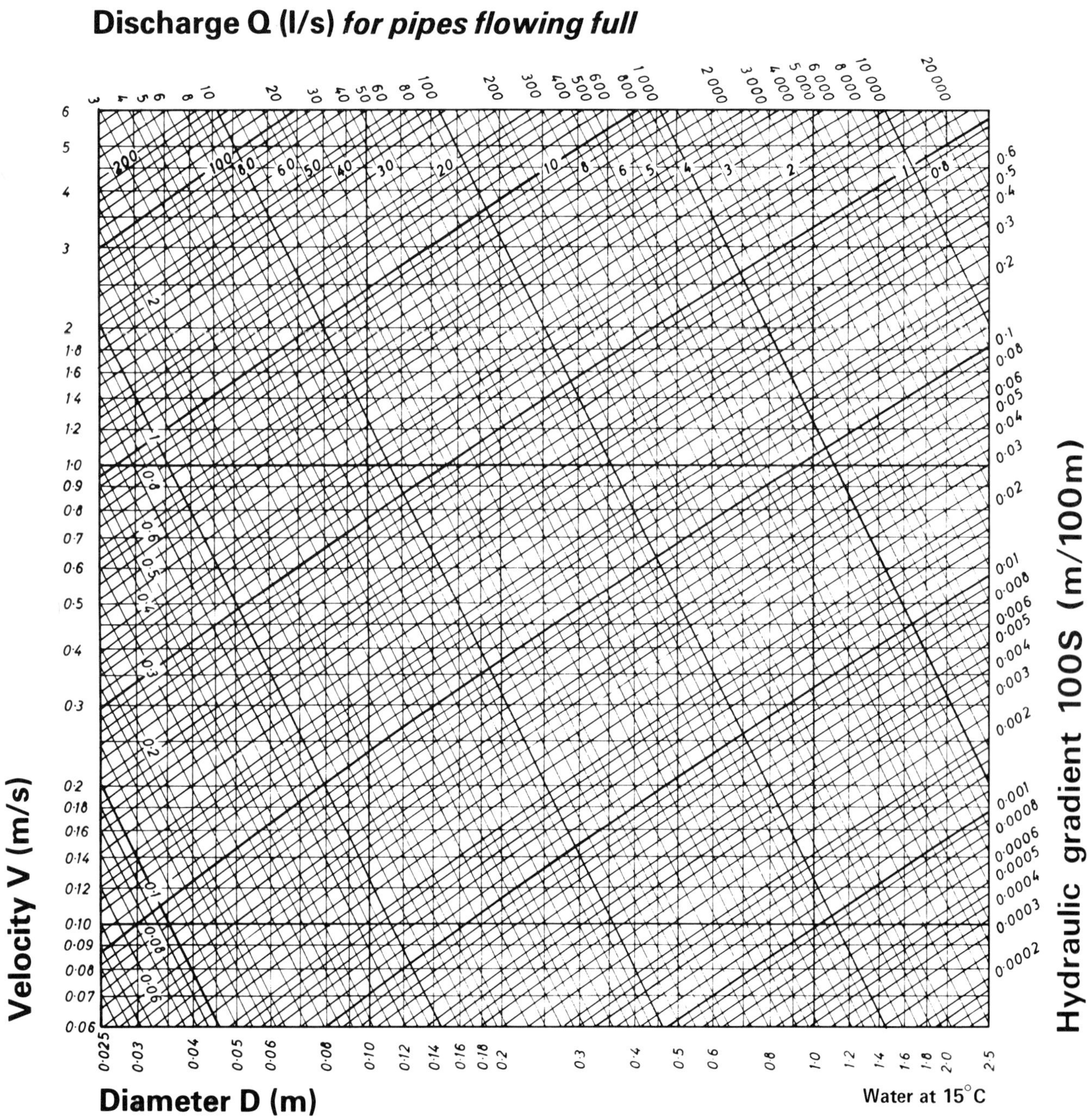

Fig 8 $k_s = 0{\cdot}6$ mm

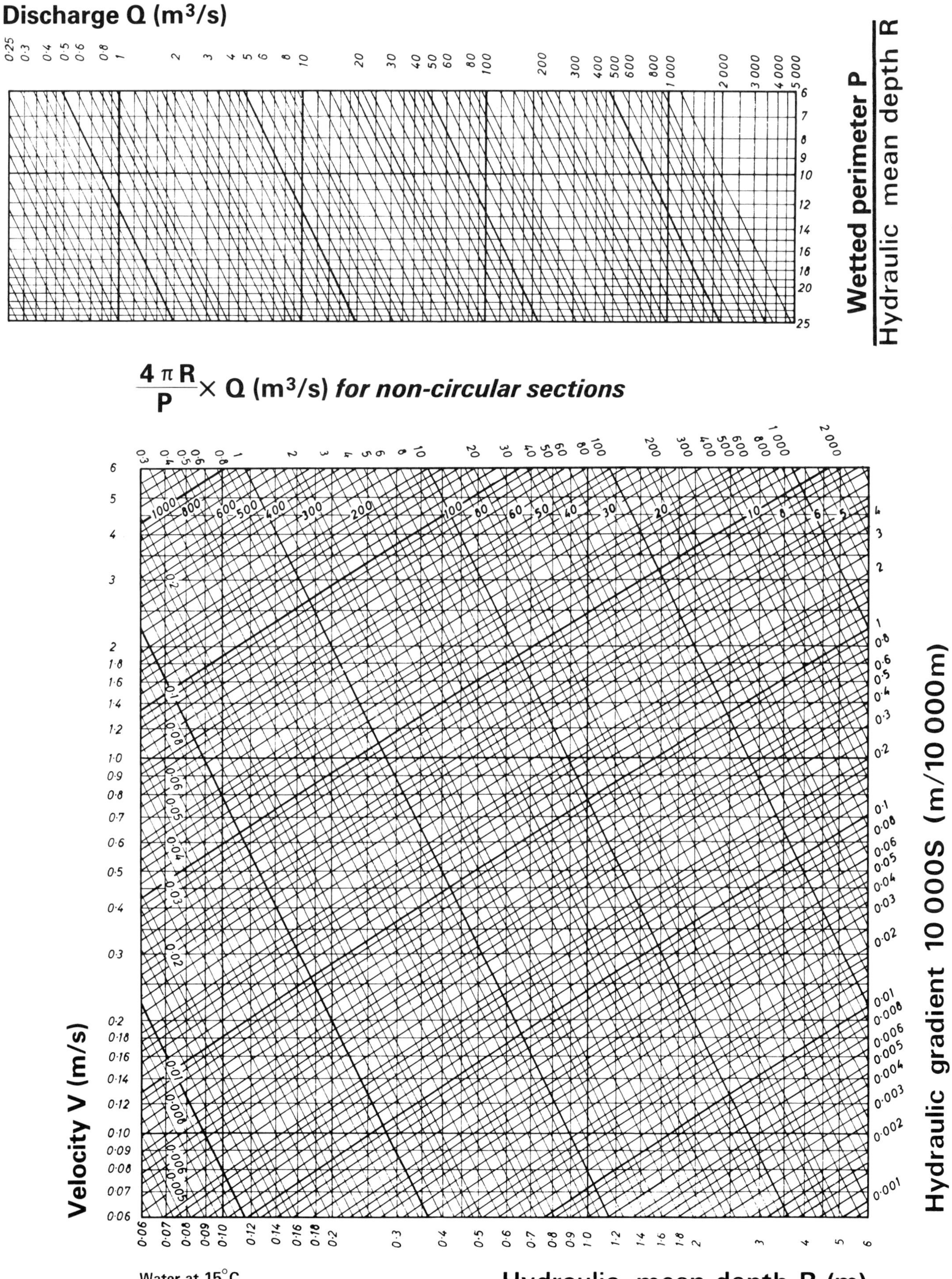

$k_s = 0{\cdot}6$ mm Fig 8

List of surfaces with roughness k_s = 1.5 mm

Good examples of	Old tuberculated water mains with moderate attack Well pointed brickwork
Normal examples of	Old tuberculated water mains with slight attack Glazed brickwork Clayware sewers slimed to about half depth; velocity, when flowing half full, approximately 0.75m/s Concrete sewers, spun or vertically cast, slimed to about half depth; velocity, when flowing half full, approximately 1.2m/s
Poor examples of	Wood stave pipes, planed plank conduits Asbestos cement sewers slimed to about half depth; velocity, when flowing half full, approximately 1.2m/s

Discharge Q (l/s) ***for pipes flowing full***

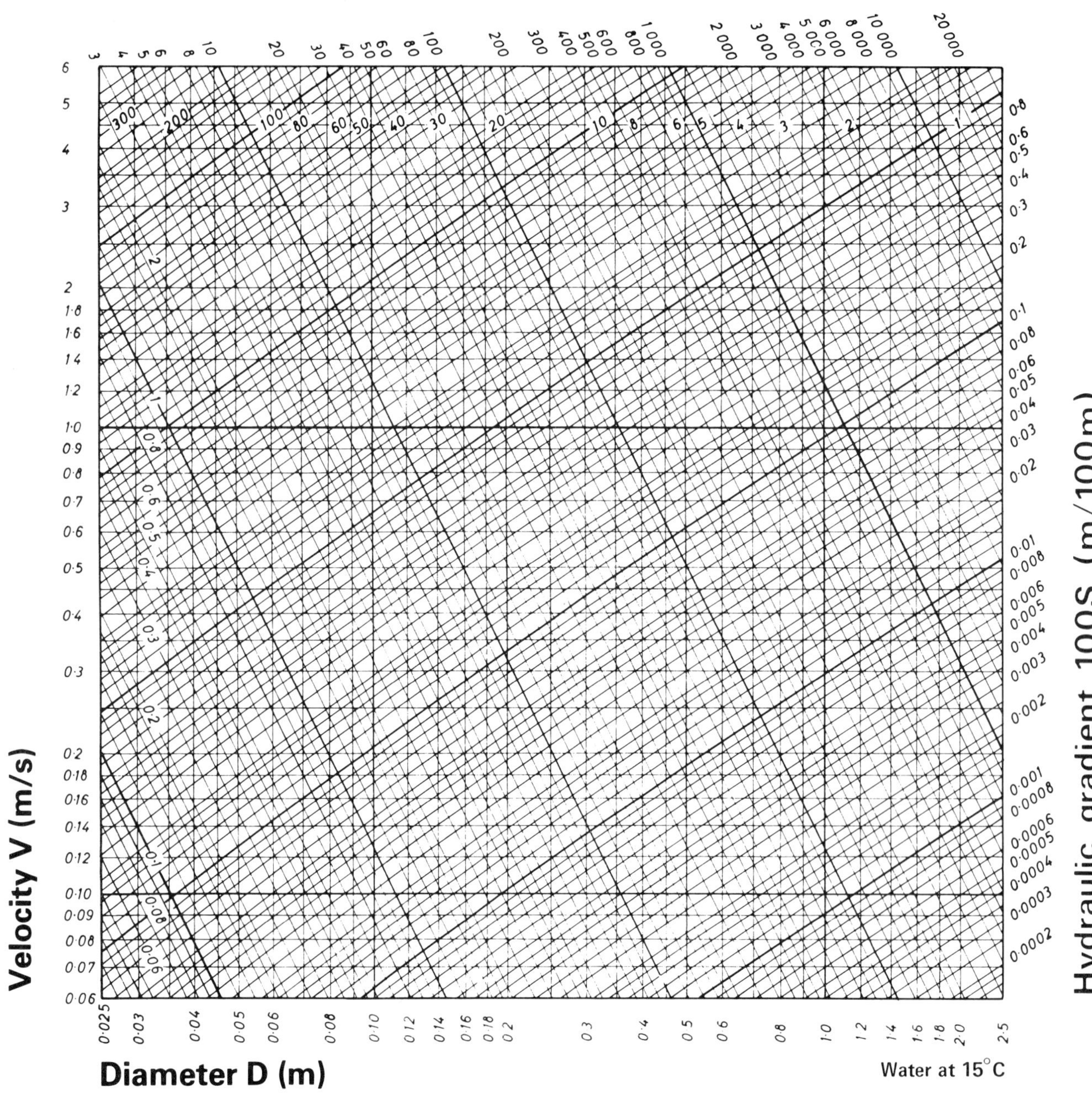

Fig 9 $k_s = 1{\cdot}5$ mm

Discharge Q (m^3/s)

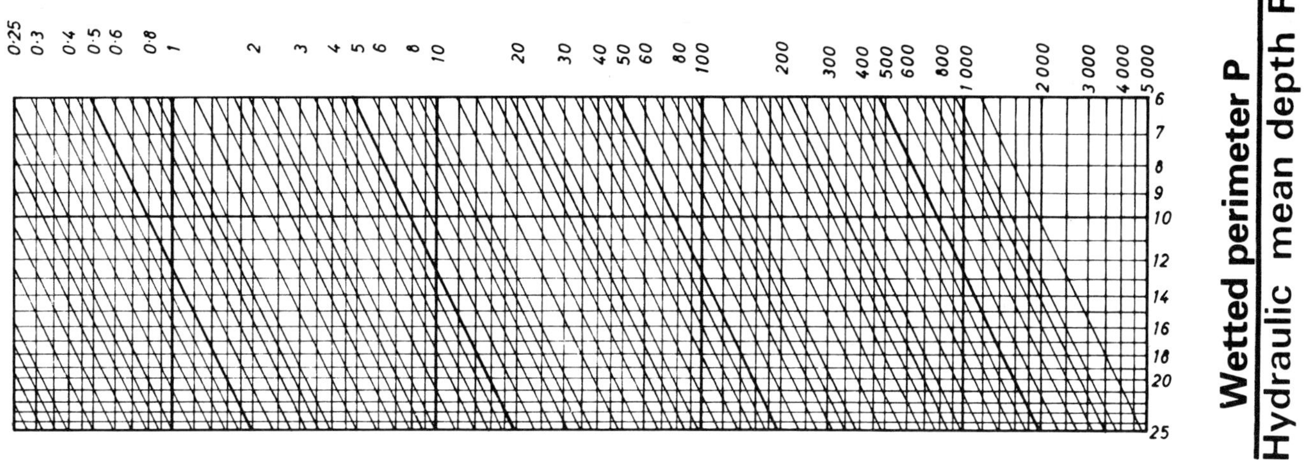

$\frac{4\pi R}{P} \times Q$ (m^3/s) *for non-circular sections*

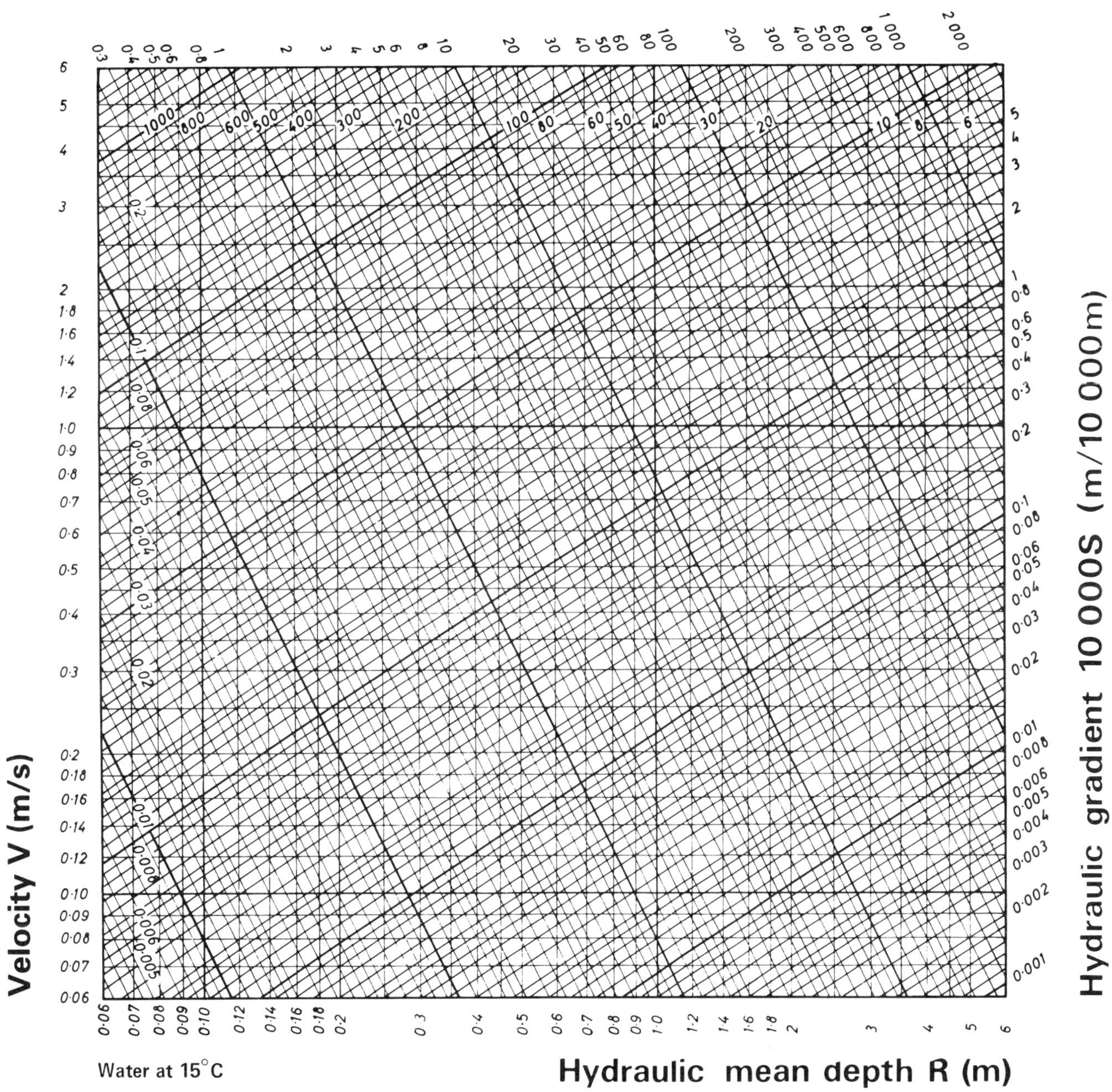

Water at 15°C

Hydraulic mean depth R (m)

$k_s = 1{\cdot}5$ mm Fig 9

List of surfaces with roughness k_s = 3.0 mm

Good examples of	
Normal examples of	Old tuberculated water mains with moderate attack Well pointed brickwork Concrete sewers, spun or vertically cast; Asbestos cement sewers — slimed to about half depth; velocity when flowing half full approximately 0.75 m/s
Poor examples of	Rusty wrought iron Old tuberculated water mains with slight attack Glazed brickwork Concrete sewers, spun or vertically cast, slimed to about half depth; velocity, when flowing half full, approximately 1.2m/s Clayware sewers slimed to about half depth; velocity, when flowing half full, approximately 0.75m/s

Discharge Q (l/s) ***for pipes flowing full***

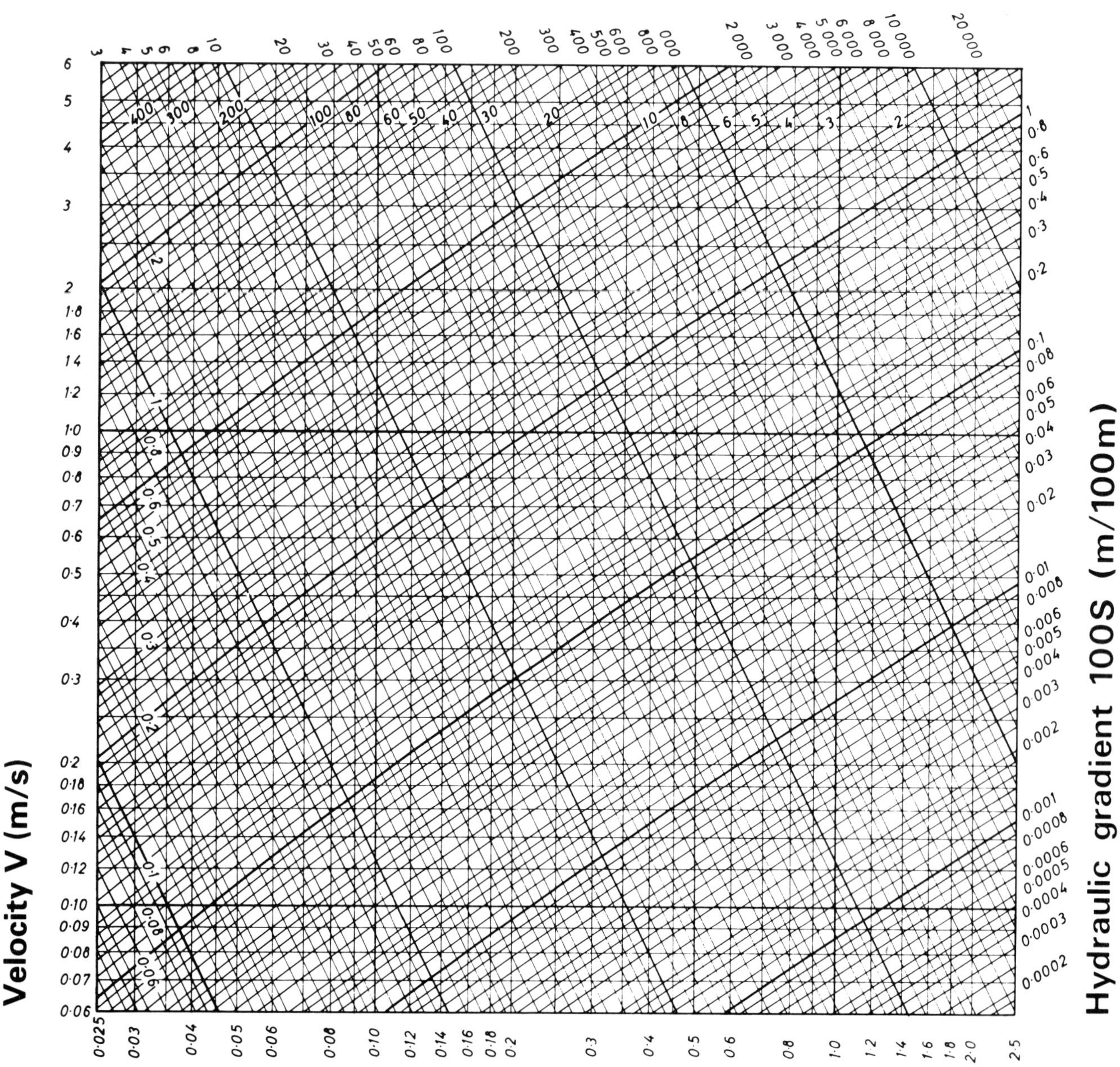

Diameter D (m)

Water at 15°C

Fig 10 $k_s = 3{\cdot}0$ mm

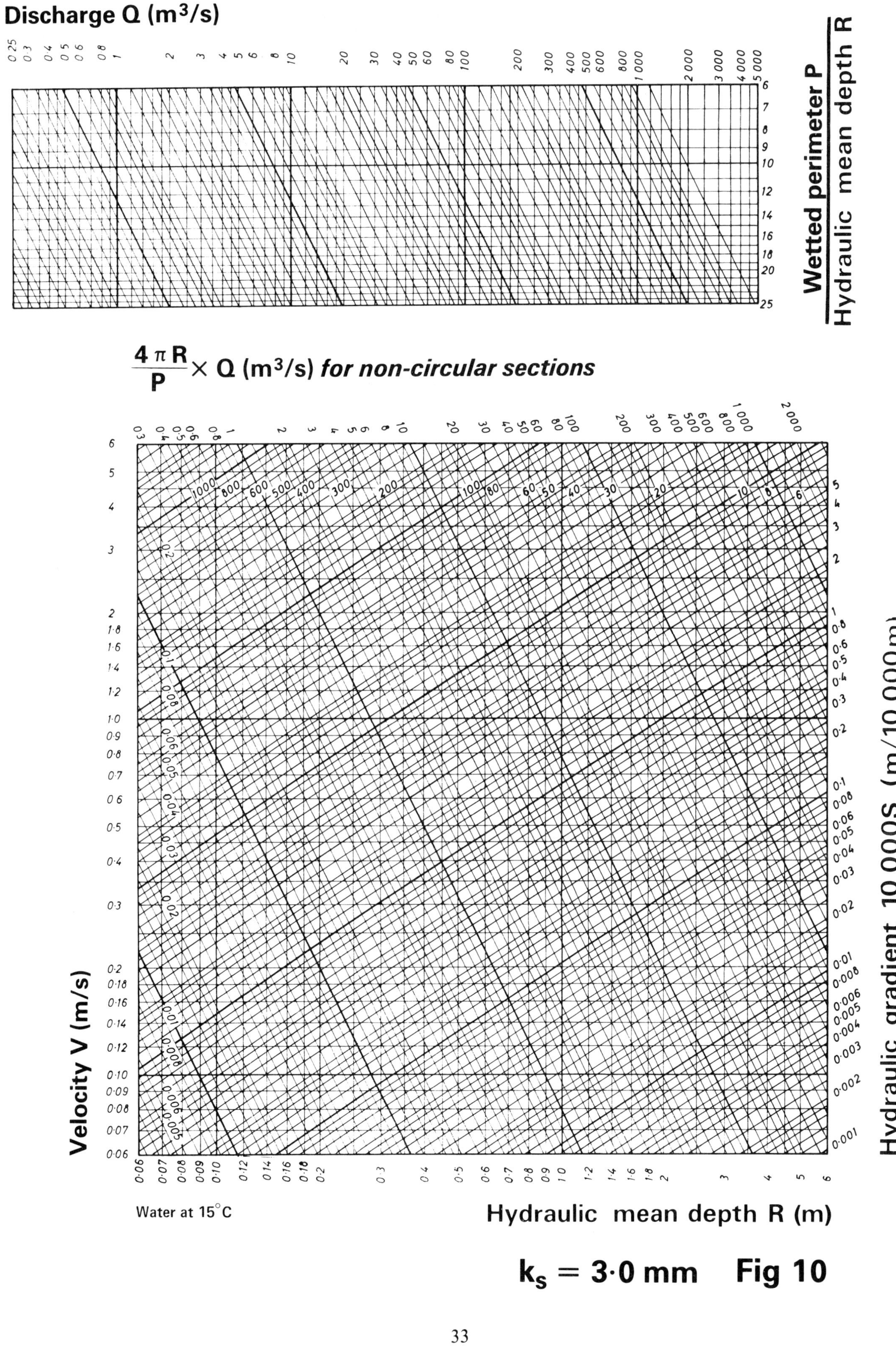

k_s = 3·0 mm Fig 10

List of surfaces with roughness k_s = 6.0 mm

Good examples of	Old tuberculated water mains with appreciable attack
Normal examples of	
Poor examples of	Old tuberculated water mains with moderate attack Well pointed brickwork Concrete sewers, spun or vertically cast } Slimed to about half depth; Asbestos cement sewers } velocity when flowing half full approximately 0.75m/s

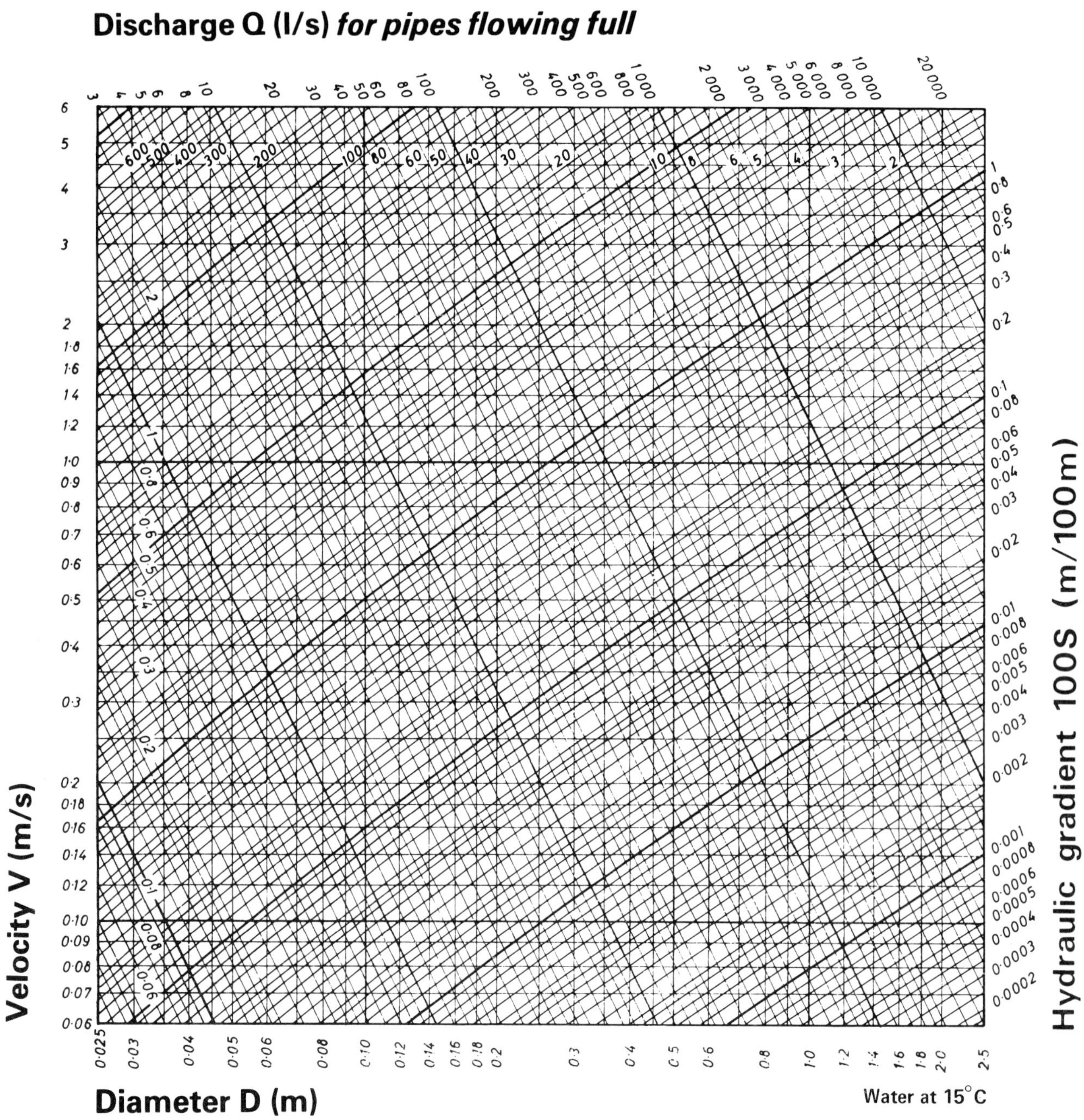

Fig 11 $k_s = 6{\cdot}0$ mm

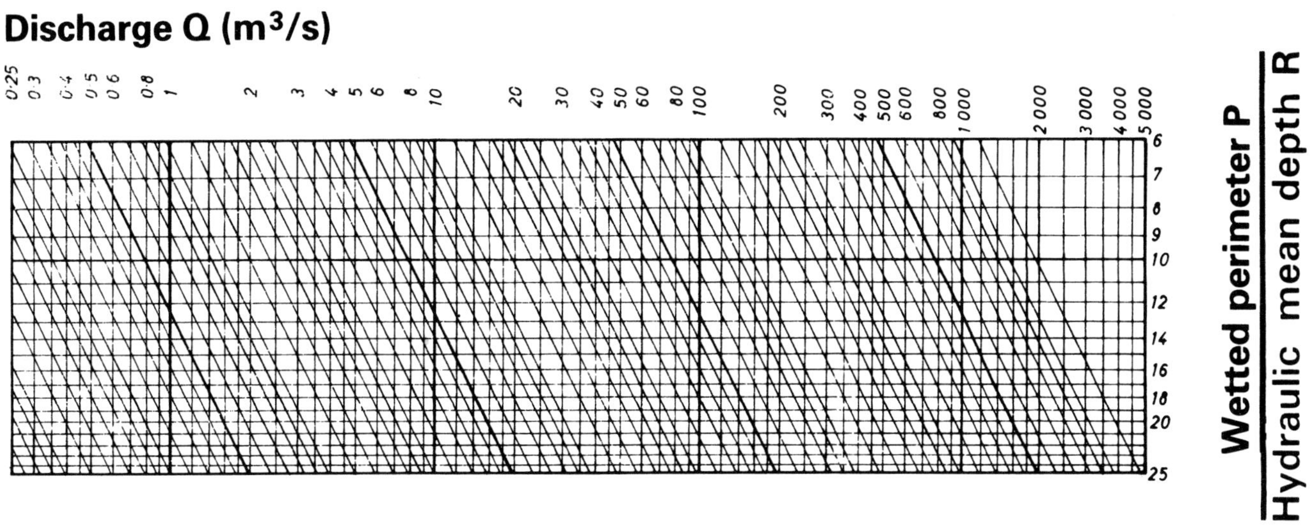

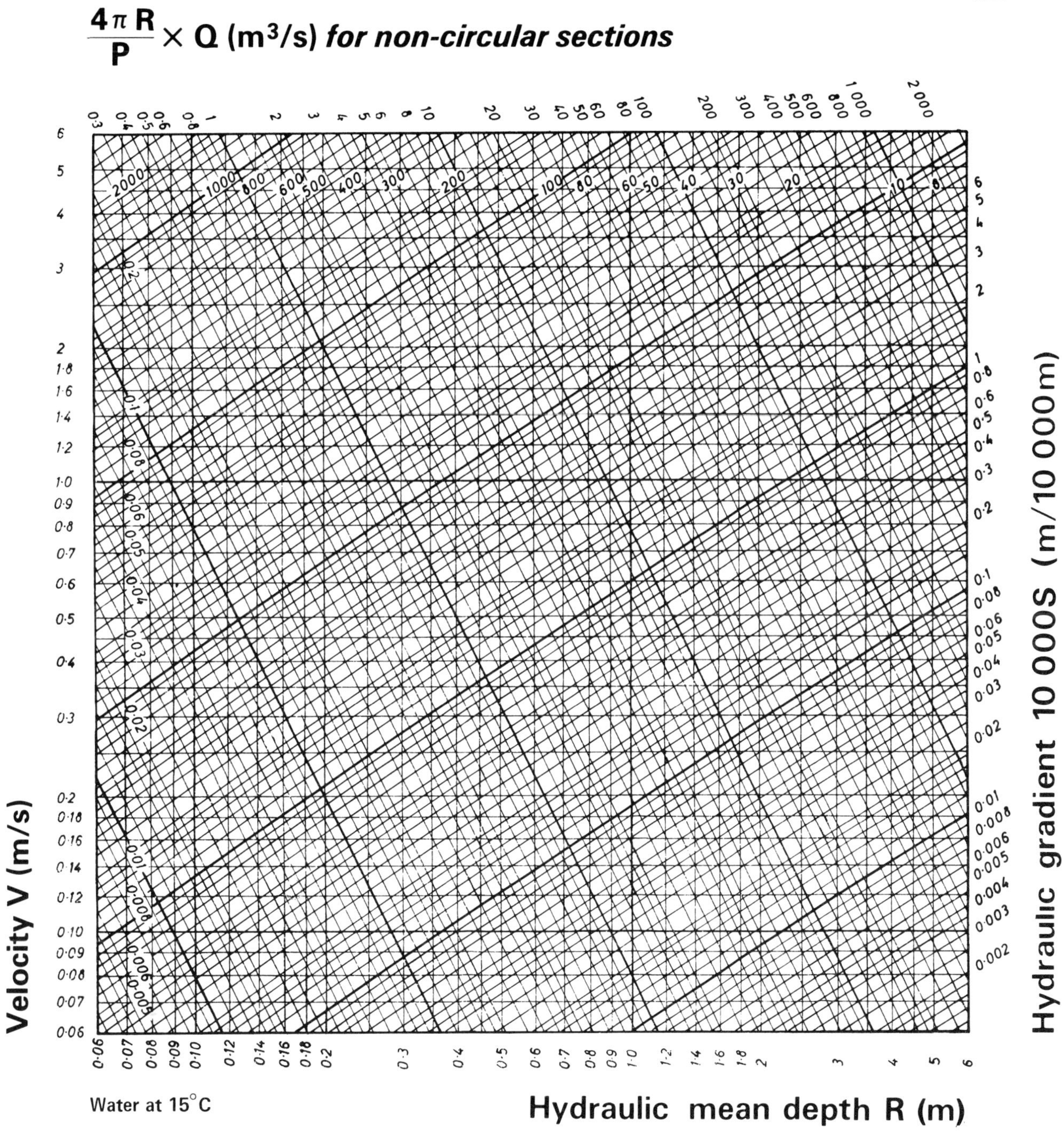

$k_s = 6{\cdot}0$ mm **Fig 11**

List of surfaces with roughness $k_s = 15$ mm

Good examples of	Old tuberculated water mains with severe attack Straight uniform artificial earth channels
Normal examples of	Old tuberculated water mains with appreciable attack Old brickwork in need of pointing
Poor examples of	

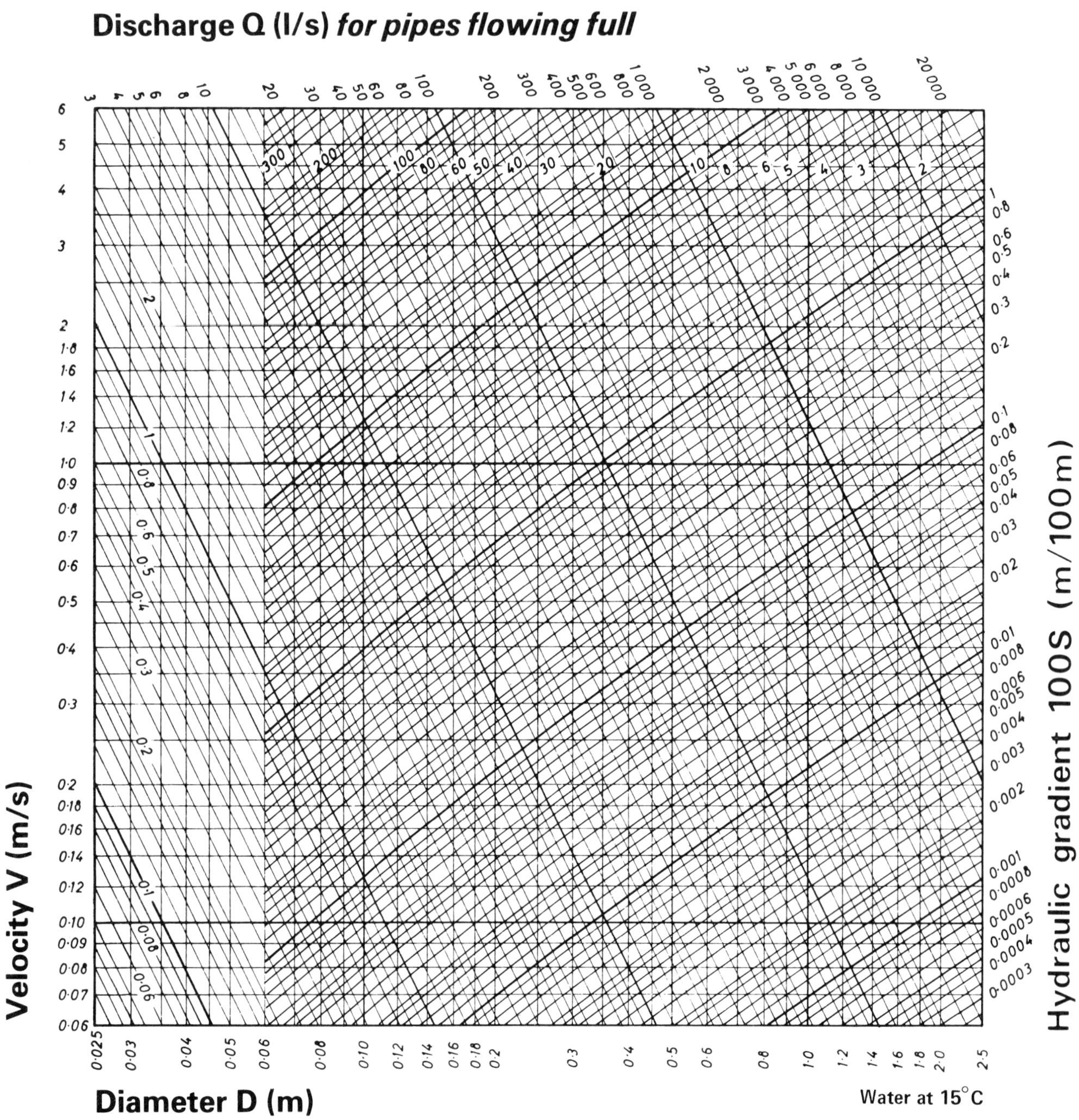

Fig 12 $k_s = 15$ mm

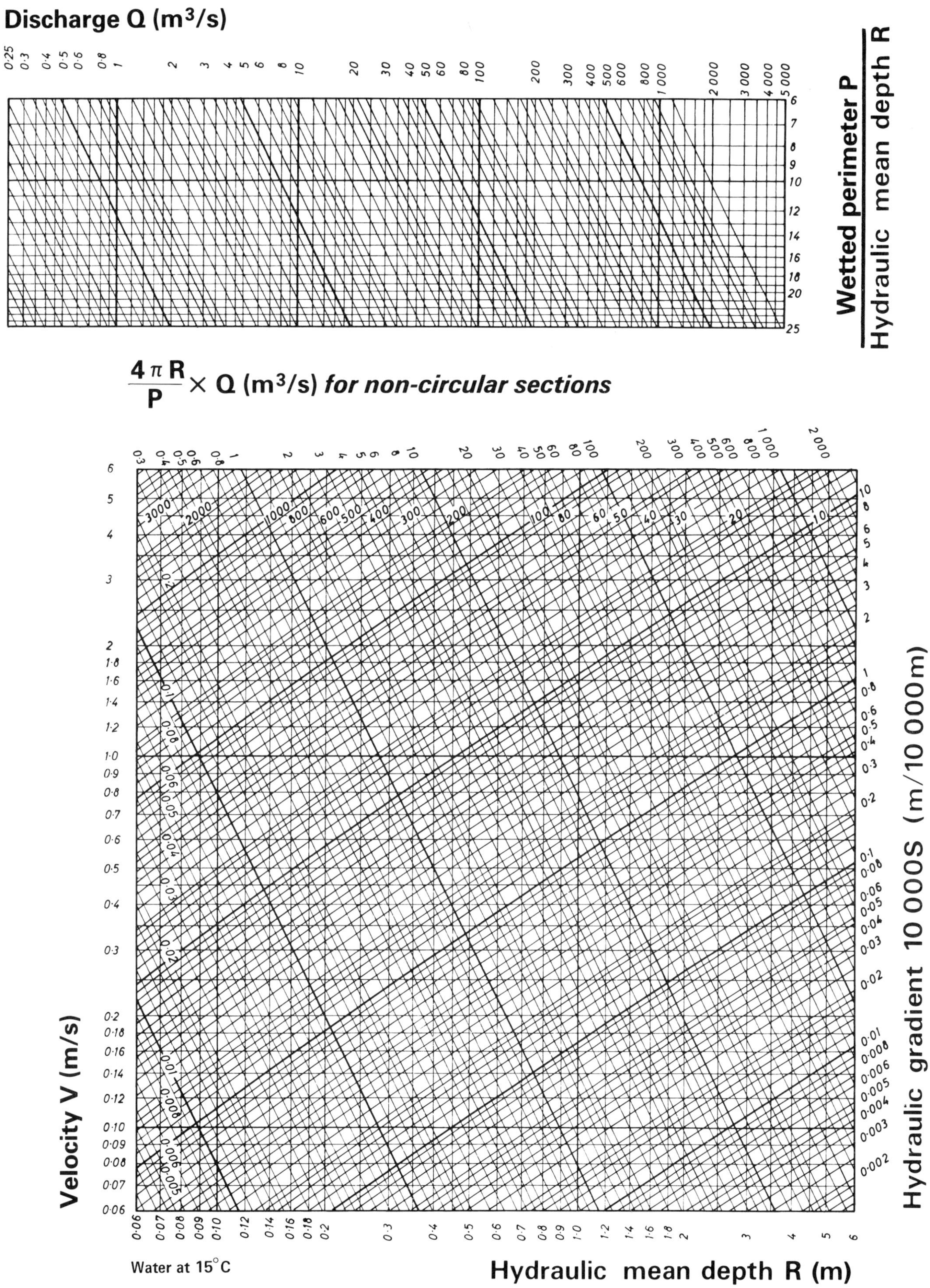

k_s = 15 mm Fig 12

List of surfaces with roughness k_s = 30 mm

Good examples of	
Normal examples of	Old tuberculated water mains with severe attack
Poor examples of	Old tuberculated water mains with appreciable attack Old brickwork in need of pointing

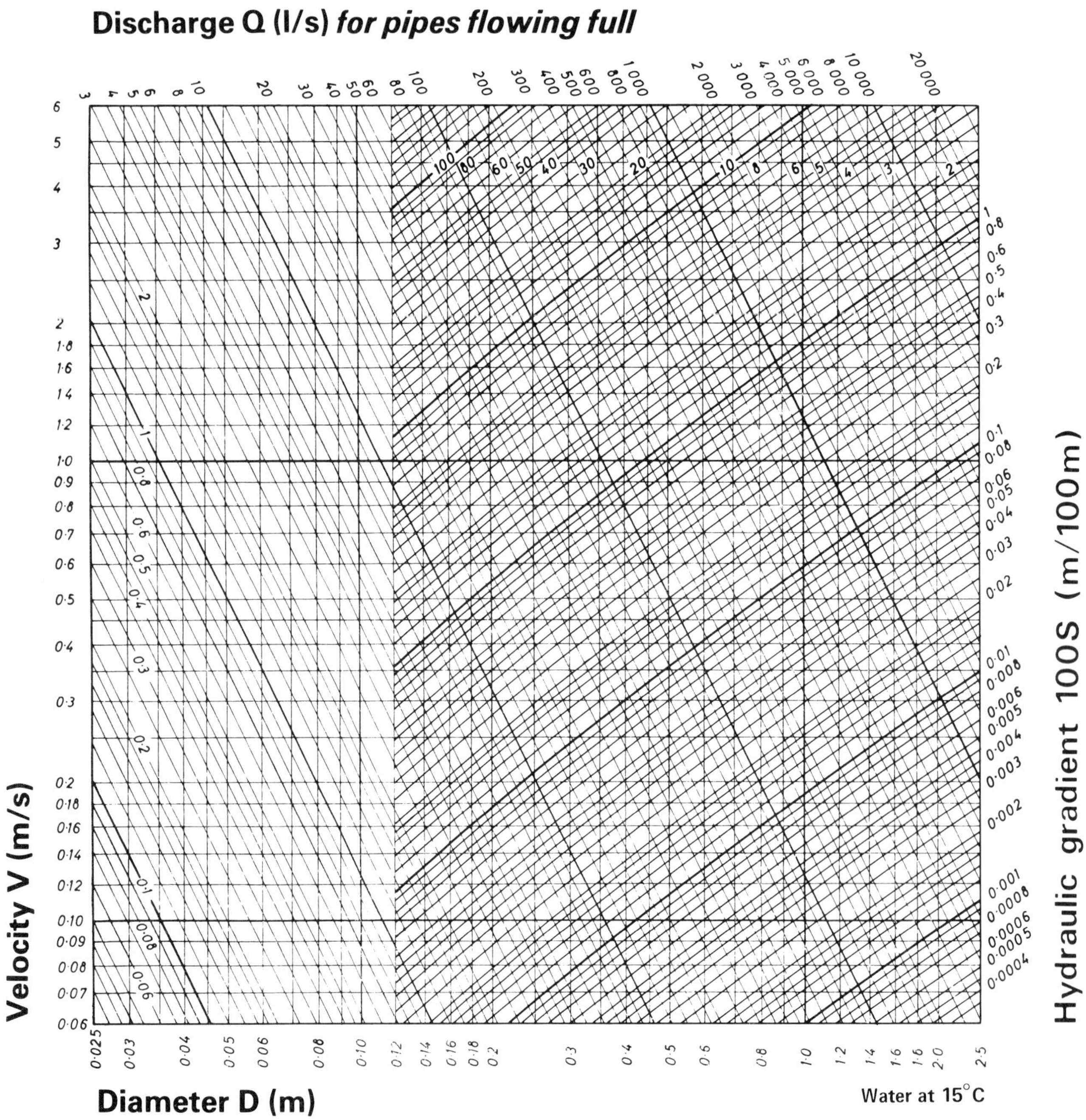

Fig 13 k_s = 30 mm

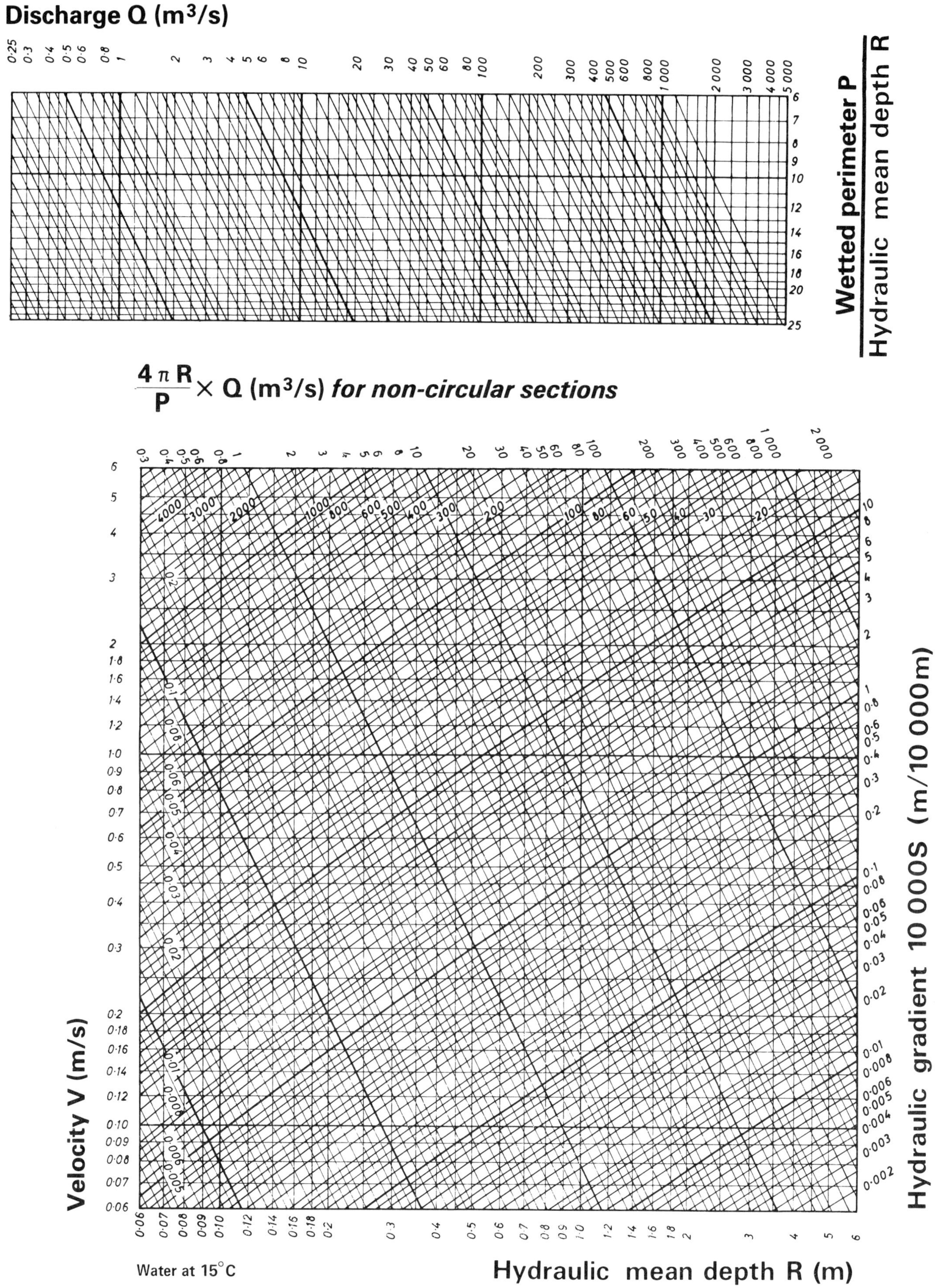

k_s = 30 mm Fig 13

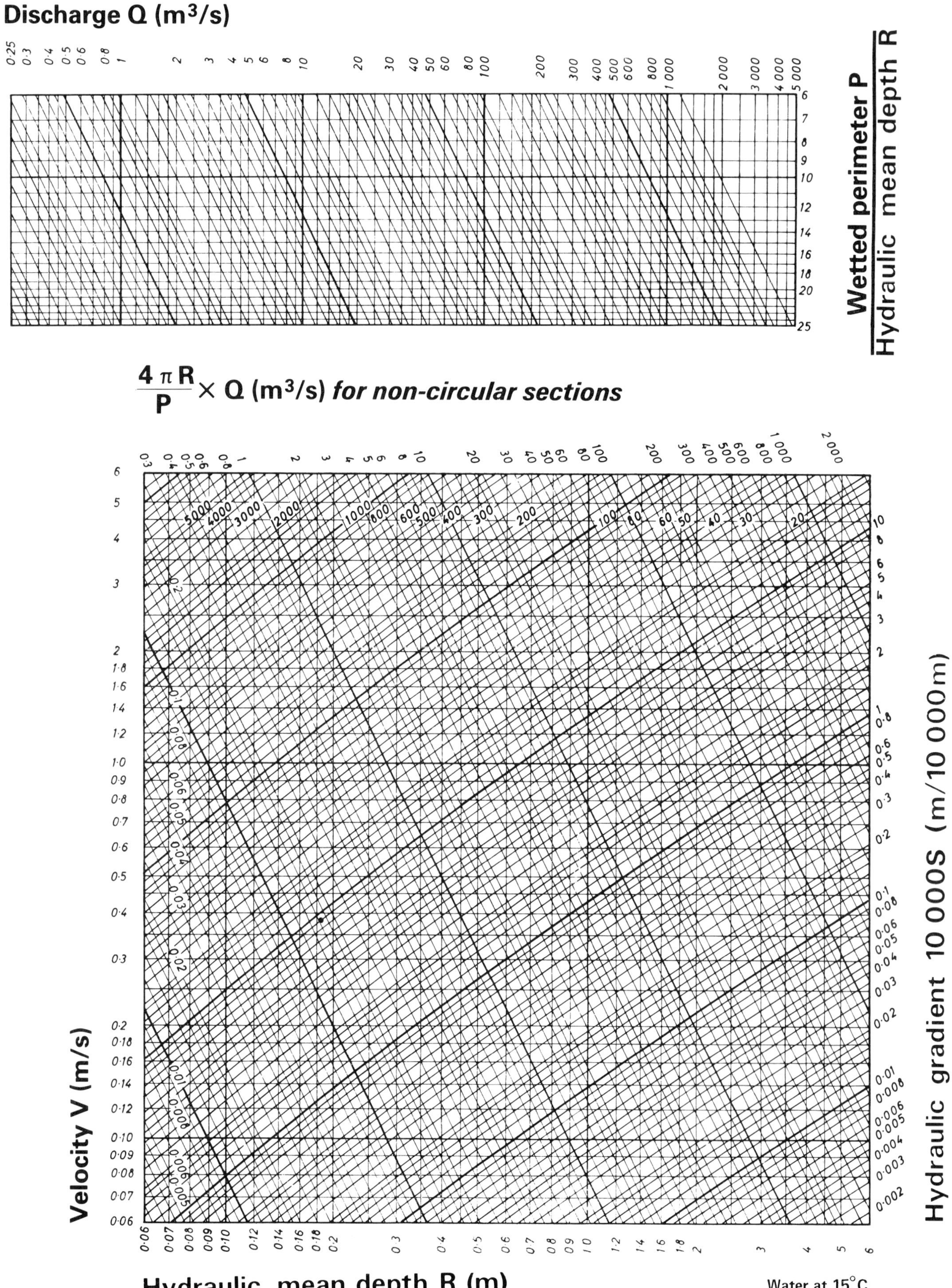

Fig 14 $k_s = 60$ mm

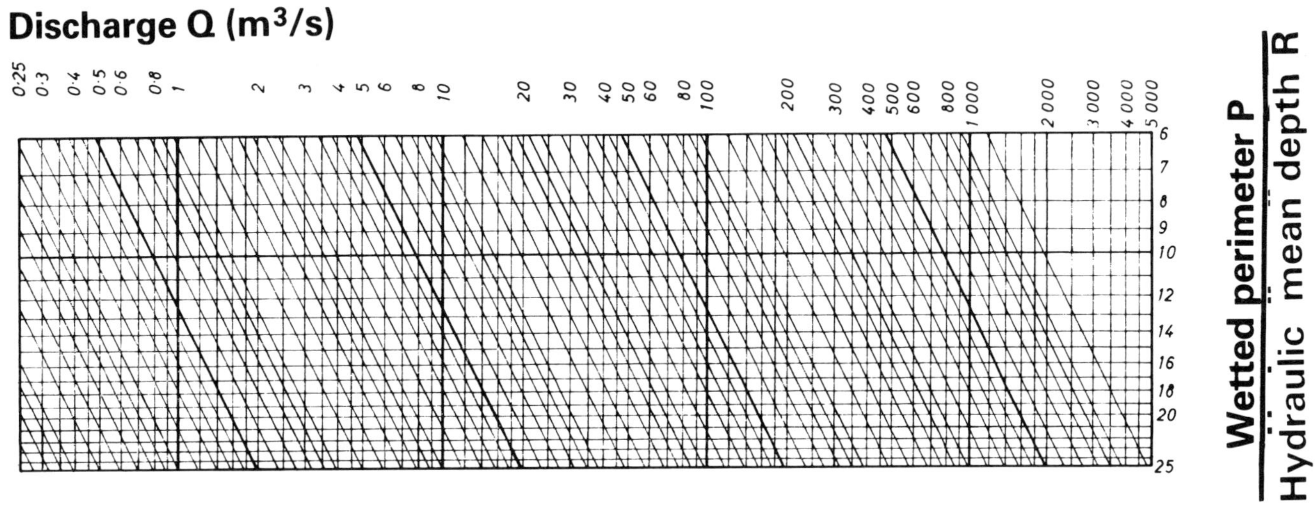

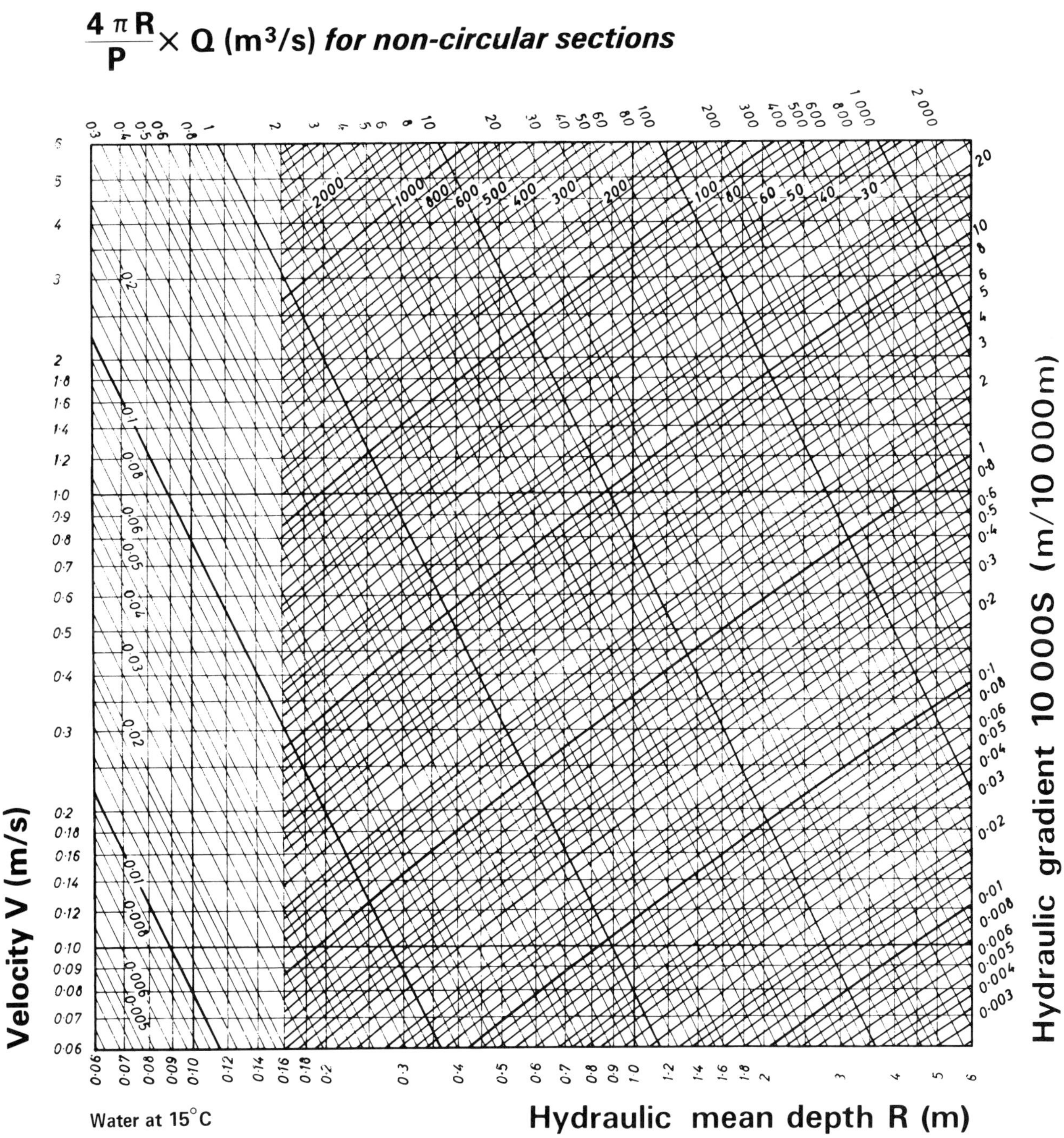

k_s = 150 mm **Fig 15**

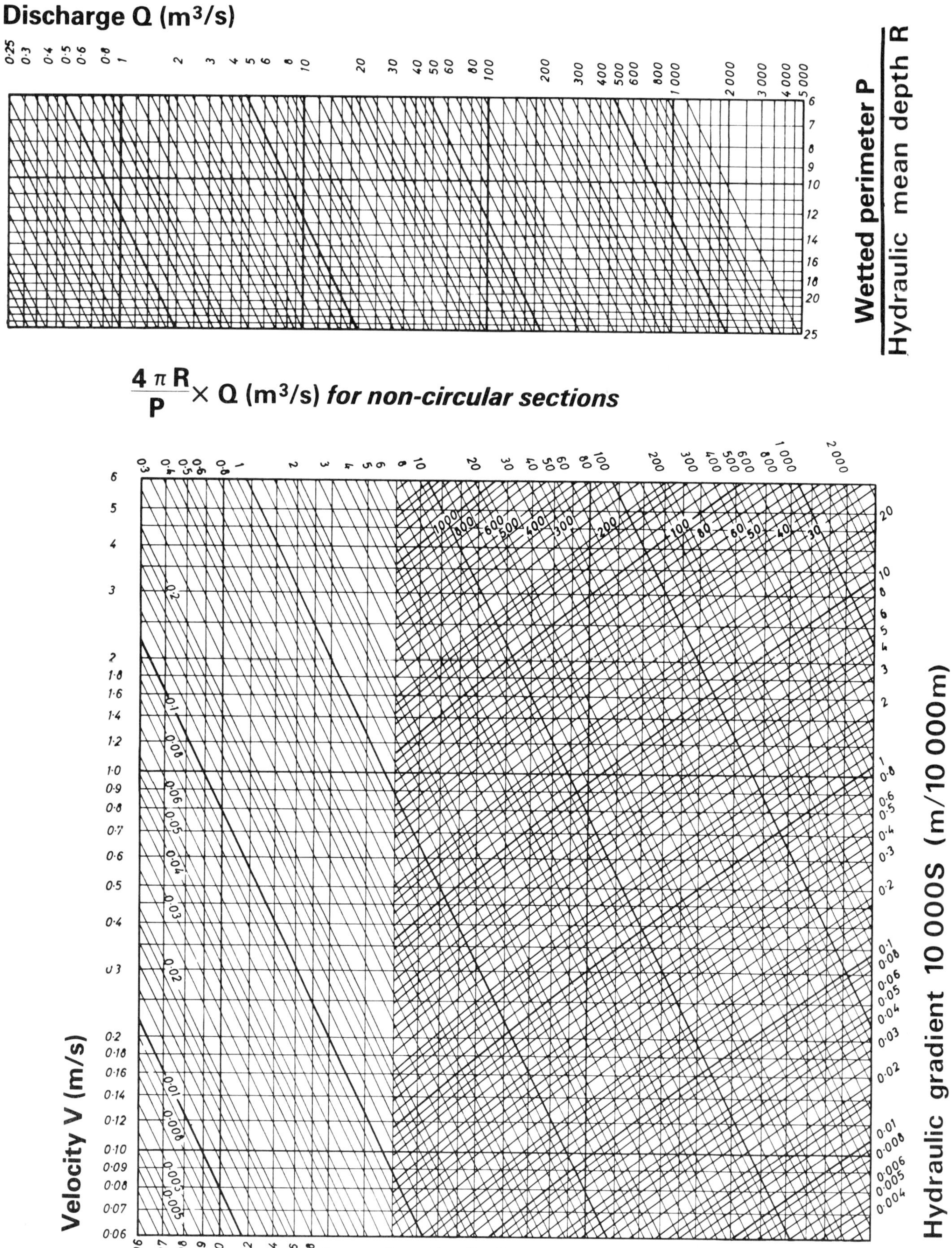

Fig 16 $k_s = 300$ mm

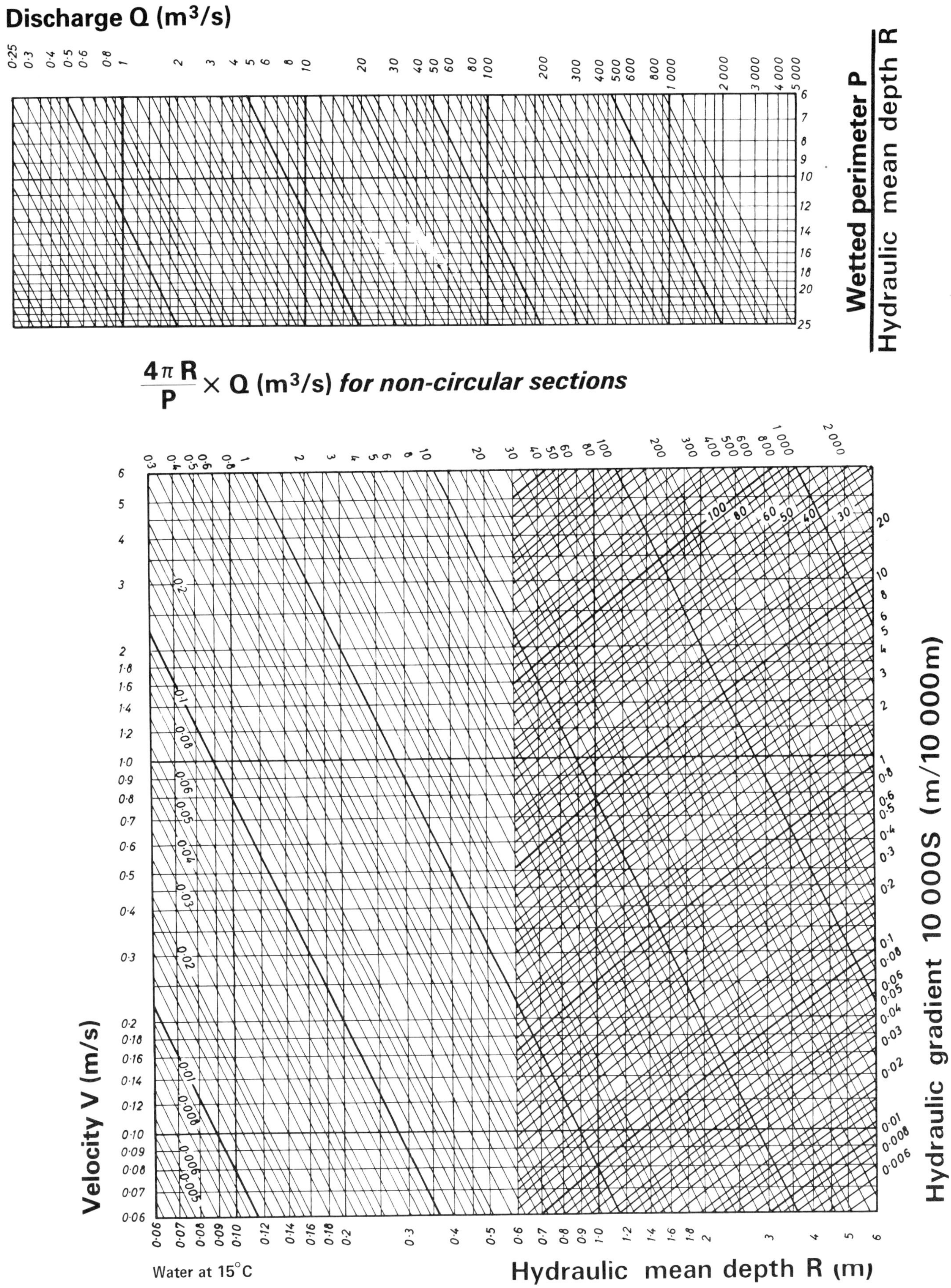

k_s = 600 mm Fig 17

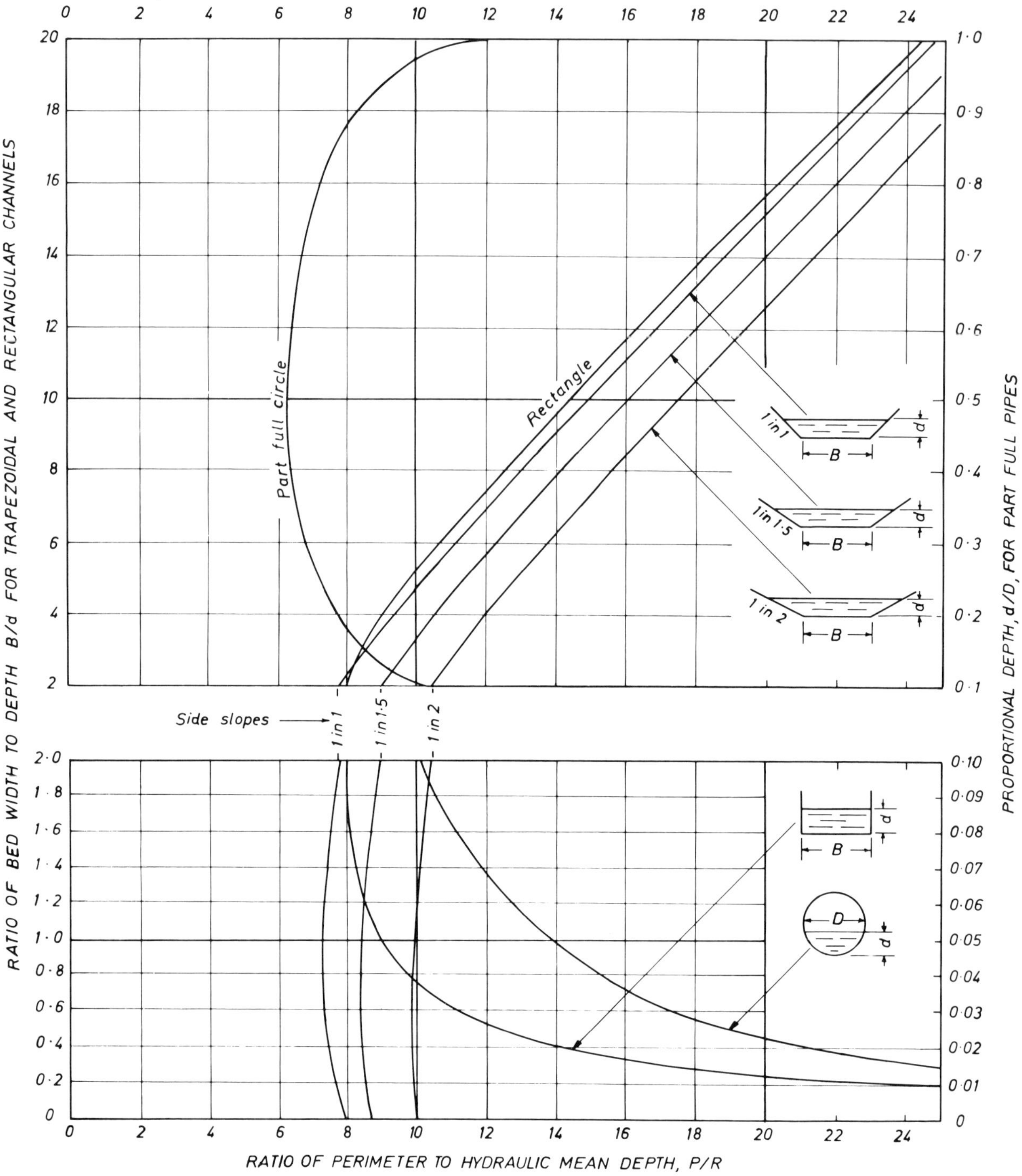

Fig 18 Shape co-efficient P/R for non-circular sections

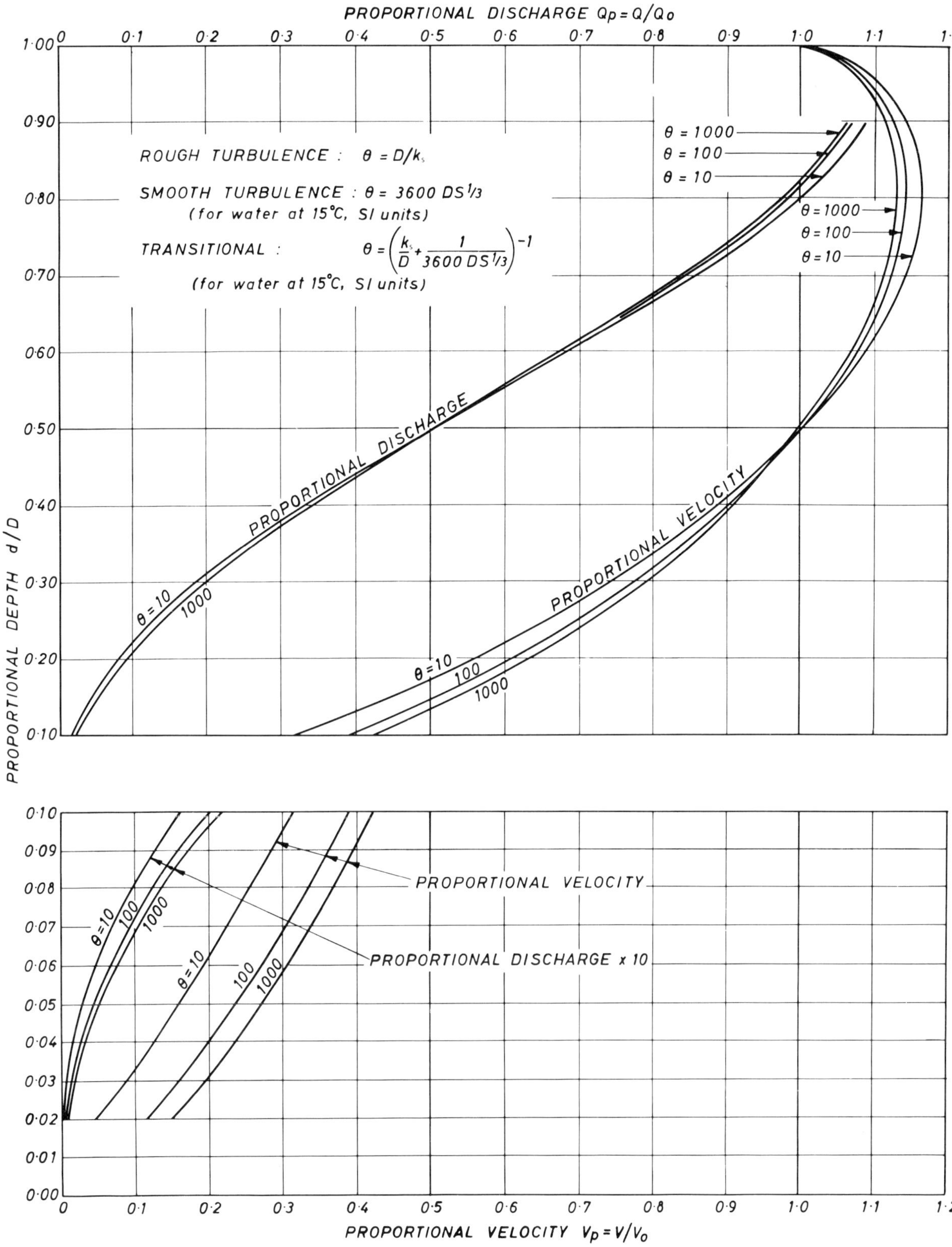

Proportional velocity and discharge in part-full circular sections **Fig 19**

Table 3 Recommended roughness values

Repeat of Table 3 from pages 12 and 13

Classification (assumed clean and new unless otherwise stated)	Suitable values of k_s (mm) Good	Normal	Poor
Smooth materials			
Drawn non-ferrous pipes of aluminium, brass, copper lead etc, and non-metallic pipes of Alkathene, glass, perspex etc	—	0.003	—
Asbestos-cement	0.015	0.03	—
Metal			
Spun bitumen or concrete lined	—	0.03	—
Wrought iron	0.03	0.06	0.15
Rusty wrought iron	0.15	0.6	3.0
Uncoated steel	0.015	0.03	0.06
Coated steel	0.03	0.06	0.15
Galvanised iron, coated cast iron	0.06	0.15	0.3
Uncoated cast iron	0.15	0.3	0.6
Tate relined pipes	0.15	0.3	0.6
Old tuberculated water mains with the following degrees of attack:			
Slight	0.6	1.5	3.0
Moderate	1.5	3.0	6.0
Appreciable	6.0	15	30
Severe	15	30	60
(Good: Up to 20 years use, Normal: 40 to 50 years use, Poor: 80 to 100 years use)			
Wood			
Wood stave pipes, planed plank conduits	0.3	0.6	1.5
Concrete			
Precast concrete pipes with 'O' ring joints	0.06	0.15	0.6
Spun precast concrete pipes with 'O' ring joints	0.06	0.15	0.3
Monolithic construction against steel forms	0.3	0.6	1.5
Monolithic construction against rough forms	0.6	1.5	—
Clayware			
Glazed or unglazed pipe:			
With sleeve joints	0.03	0.06	0.15
With spigot and socket joints and 'O' ring seals — dia < 150mm	—	0.03	—
With spigot and socket joints and 'O' ring seals — dia > 150mm	—	0.06	—
Pitch fibre (lower value refers to full bore flow)	0.003	0.03	—
Glass fibre	—	0.06	—
uPVC			
With chemically cemented joints	—	0.03	—
With spigot and socket joints, 'O' ring seals at 6 to 9 metre intervals	—	0.06	—
Brickwork			
Glazed	0.6	1.5	3.0
Well pointed	1.5	3.0	6.0
Old, in need of pointing	—	15	30

Table 3 (Cont'd)

Classification (assumed clean and new unless otherwise stated)	Suitable values of k_s (mm) Good	Normal	Poor
Slimed sewers*			
Sewers slimed to about half depth; velocity, when flowing half full, approximately 0.75m/s:			
Concrete, spun or vertically cast	—	3.0	6.0
Asbestos cement	—	3.0	6.0
Clayware	—	1.5	3.0
uPVC	—	0.6	1.5
Sewers slimed to about half depth; velocity, when flowing half full, approximately 1.2m/s:			
Concrete, spun or vertically cast	—	1.5	3.0
Asbestos cement	—	0.6	1.5
Clayware	—	0.3	0.6
uPVC	—	0.15	0.3
Sewer rising mains All materials, operating as follows			
Mean velocity 1m/s	0.15	0.3	0.6
Mean velocity 1.5m/s	0.06	0.15	0.30
Mean velocity 2m/s	0.03	0.06	0.15
Unlined rock tunnels			
Granite and other homogeneous rocks	60	150	300
Diagonally bedded slates (values to be used with *design* diameter)	—	300	600
Earth channels			
Straight uniform artificial channels	15	60	150
Straight natural channels, free from shoals, boulders and weeds	150	300	600

* The roughness of a slimed sewer varies considerably during any year. The normal value is that roughness which is exceeded for approximately half of the time. The poor value is that which is exceeded, generally on a continuous basis, for one month of the year. The value of k_s should be interpolated for velocities between 0.75 and 1.2m/s.

Hydraulics Research, Wallingford, is an independent company limited by guarantee registered in the UK as Hydraulics Research Station Limited. It was formed in 1982 to take over and extend the activities of the British Government's highly acclaimed Hydraulics Research Station, which for 35 years so successfully served the civil engineering industry.

We enjoy close ties with British civil engineering consultants and have extensive international experience in all branches of civil engineering hydraulics. We are always pleased to discuss particular problems with potential clients and to develop appropriate solutions using the wide range of sophisticated techniques derived from our basic research studies. We can advise on the setting up of specialist laboratories and on the collection of field data. Our specialists are available at short notice to make site visits. In our engineering studies we frequently make use of computational or physical models or undertake field investigations. Sometimes all three may be necessary. Often a site visit or desk study is sufficient if only a limited investigation is required.

Organizations in the UK and overseas for whom we regularly carry out investigations include consulting engineers and contractors, industrial concerns, public authorities and government agencies.

ISBN: 0 946466 02 5